W0106190

Patric Nisbet · Wladyslaw Gedroyc · Sheila Rankin

Case Studies in Diagnostic Imaging

Film Interpretation for Postgraduate Examinations

With 80 Figures

Springer-Verlag
London Berlin Heidelberg New York
Paris Tokyo

Patric Nisbet, MSc, MD, MRCP, FRCR
Senior Registrar in Radiology, X-Ray Department, Guy's Hospital,
St Thomas Street, London SE1 9RT, UK

Wladyslaw Gedroyc, MB, BS, MRCP, FRCR
Senior Registrar in Radiology, X-Ray Department, Guy's Hospital,
St Thomas Street, London SE1 9RT, UK

Sheila Rankin, MB, BS, DMRD, FRCR
Consultant Radiologist, CT Scanning Unit, X-Ray Department,
Guy's Hospital, St Thomas Street, London SE1 9RT, UK

ISBN-13: 978-3-540-19500-9   e-ISBN-13: 978-1-4471-1601-1
DOI: 10.1007/978-1-4471-1601-1

British Library Cataloguing in Publication Data
Nisbet, Patric
Case studies in diagnostic imaging.
1. Imaging systems in medicine 2. Diagnosis
I. Title   II. Gedroyc, Wladyslaw
III. Rankin, Sheila
616.07'54   R857.06
ISBN-13: 978-3-540-19500-9

Library of Congress Cataloging-in-Publication Data
Nisbet, A. P. (Andrew Patric), 1951–
Case studies in diagnostic imaging.
Includes bibliographies. 1. Diagnostic imaging—Case studies. I. Gedroyc,
W. M. W. (Wladyslaw Michal W.), 1954–   . II. Rankin, S. C. (Sheila Campbell),
1948–   . III. Title. [DNLM: 1. Nuclear Magnetic Resonance— examination
questions. 2. Radiography—examination questions. 3. Radionuclide Imaging—
examination questions. 4. Tomography, Emission Computed—examination
questions. 5. Tomography, X-Ray Computed—examination questions.
6. Ultrasonic Diagnosis—examination questions. WN 18 N723c]
RC78.7.D53N57 1987 616.07'57 87–16298
ISBN-13: 978-3-540-19500-9

Filmset by Wilmaset, Birkenhead, Wirral

2128/3916–543210

Preface

Imaging now plays an integral part in most diagnostic pathways. A familiarity with plain-film abnormalities and the more specialised modalities such as computerised tomography, ultrasound, nuclear medicine and magnetic resonance is an important part of the clinician's "medical knowledge", and a sound grasp of the subject is expected in postgraduate examinations.

This book is primarily intended as a study guide for film interpretation in postgraduate examinations, especially the MRCP and FRCR exams.

The layout of the questions follows the format of the examination of the Royal College of Radiologists. On right-hand pages brief clinical details and one or more imaging examples are presented for the candidate to analyse and report. In each case specimen answers with comments and, where necessary, further illustrations, are shown on the following left-hand page.

This text cannot be comprehensive, but it should form a foundation for future study.

Finally, we are very grateful to our many colleagues who have provided additional material for this book.

London, 1987

Patric Nisbet
Wladyslaw Gedroyc
Sheila Rankin

Introduction

Guidance for FRCR Candidates

Since April 1984 the Final Fellowship examination of the Royal College of Radiologists has included a new section: the Film Viewing Session. One hour is allotted to this session, with eight candidates being examined at one time using identical sets of cases. Each candidate is given eight packets of films and each packet may contain up to three films. Relevant clinical data is included with each individual case.

Reports are expected to include observations, interpretations, conclusions and recommendations for further investigations where appropriate. The examination guidance notes emphasise that reports must be brief and relevant.

The cases are of differing complexity and cover all imaging modalities. To be successful in the Final Fellowship examination, the candidate must obtain a pass standard in this session.

Our experience has shown that organised practice in the film viewing session format is of value to examination candidates. Most radiologists report films verbally, either directly to a secretary or into a tape recorder. Producing accurate, legible, handwritten reports within a set time limit is an unfamiliar task which requires practice.

The time available must be organised to produce an adequate answer for each case, and this book offers ten sample one-hour examinations for readers to attempt. Specimen answers are provided after each case, with comments and supplementary films to clarify the diagnosis where necessary. In assessing the answers, the reader should be aware of the closed marking system used by examiners, grading answers P+, P, P− and F (P = pass, F = fail). In this system a sensible, well-analysed answer should attain a P−, even if the correct diagnosis does not head the list produced by the candidate; similarly a superb answer can only achieve a P+. It is thus of great importance to devote an equal time to each of the eight questions, rather than to produce seven high-quality answers and not answer the eighth question. It should also be remembered that variable weighting is applied to questions according to their complexity, so that a perfect answer to one question may count less than a moderate answer to another.

The specimen answers are not perfect, but in this type of examination there is no single correct answer and we have only attempted to show what can be done within the time available. Similarly, the illustrations are not necessarily of the highest quality, but in the actual examination all the films are copies and consequently have the failings of copy films. If the text illustrations also show these failings we feel that the examination conditions have been closely simulated.

However, if the candidate allocates time wisely, describes the important positive and negative findings; lists the differential diagnoses with an indication of likelihood, and follows with a comment on useful further investigations, the examiners are likely to be satisfied.

Contents

Examination 5

Examination 6

Examination 7

Examination 8

Examination 9

Examination 10

Examination 1

Examination 1

Clinical Data

A 55-year-old male with abdominal pain.

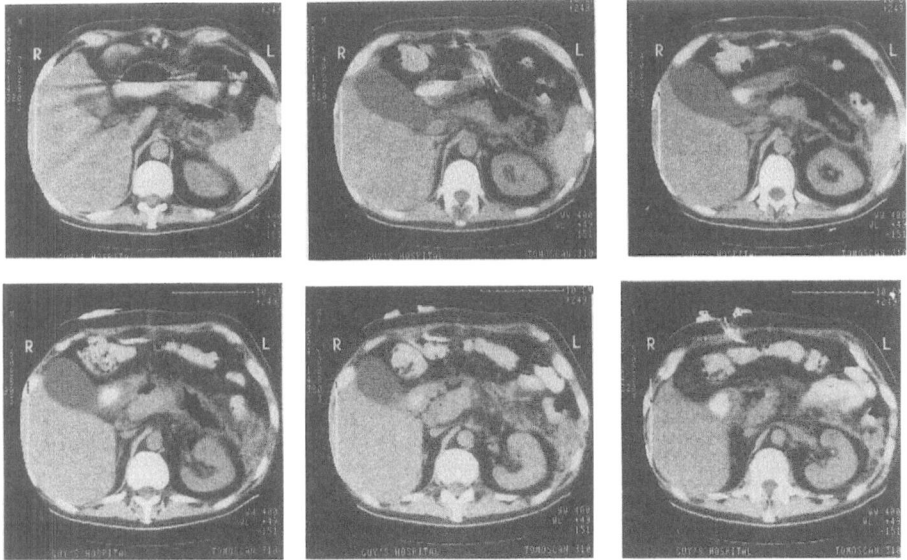

Fig. 1.1a

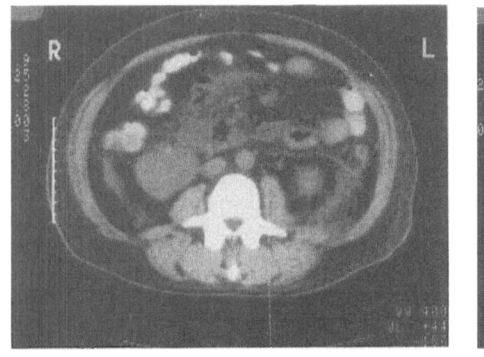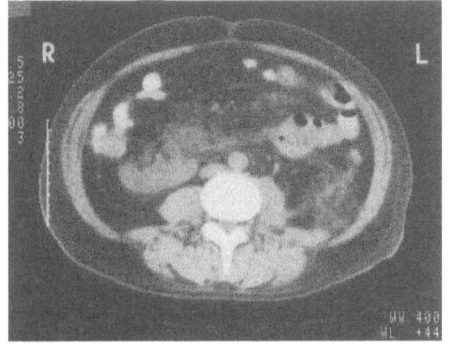

Fig. 1.1b

Pancreatitis

Case Report

CT Upper Abdomen: Gas is visible in an enlarged pancreatic bed and the pancreatic outlines are difficult to define. An external drain runs into the pancreatic bed.

The wall of the descending duodenal loop wall is thickened. The mesenteric fat anterior to the pancreas is abnormally dense, suggesting fat necrosis. There is marked thickening of the left anterior pararenal fascia in continuity with the pancreas.

Diagnosis

The appearances are those of acute necrotising pancreatitis, and the extensive gas within the pancreatic bed suggests superimposed abscess formation.

Clinical Data

A 25-year-old female presenting with jaundice.

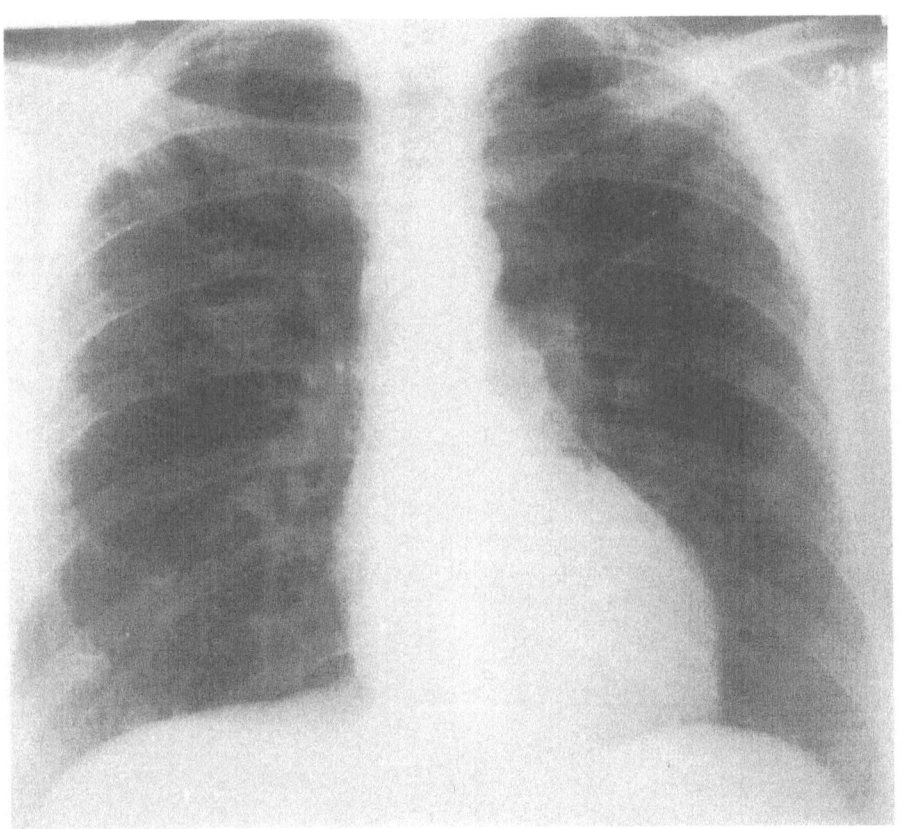

Fig. 1.2

5

Multiple Abscesses

Case Report

PA Chest: Multiple, variable-sized nodules, 1–4 cm in diameter, are present in all lung zones. Several nodules show cavitation and at least one contains an air–fluid level. The hila are normal, as are the visible bones. The heart is not enlarged.

Differential Diagnosis

1. Multiple abscesses; these are more likely in this young age group. Is there known sepsis or evidence of drug abuse?
2. Metastases.
3. Pulmonary infarcts; the distribution is atypical for this, however.
4. Rheumatoid nodules; less likely, particularly in the absence of pleural effusions.
5. Cavitating sarcoid is possible, but is rare.

Clinical Data

A 25-year-old male with dysphagia.

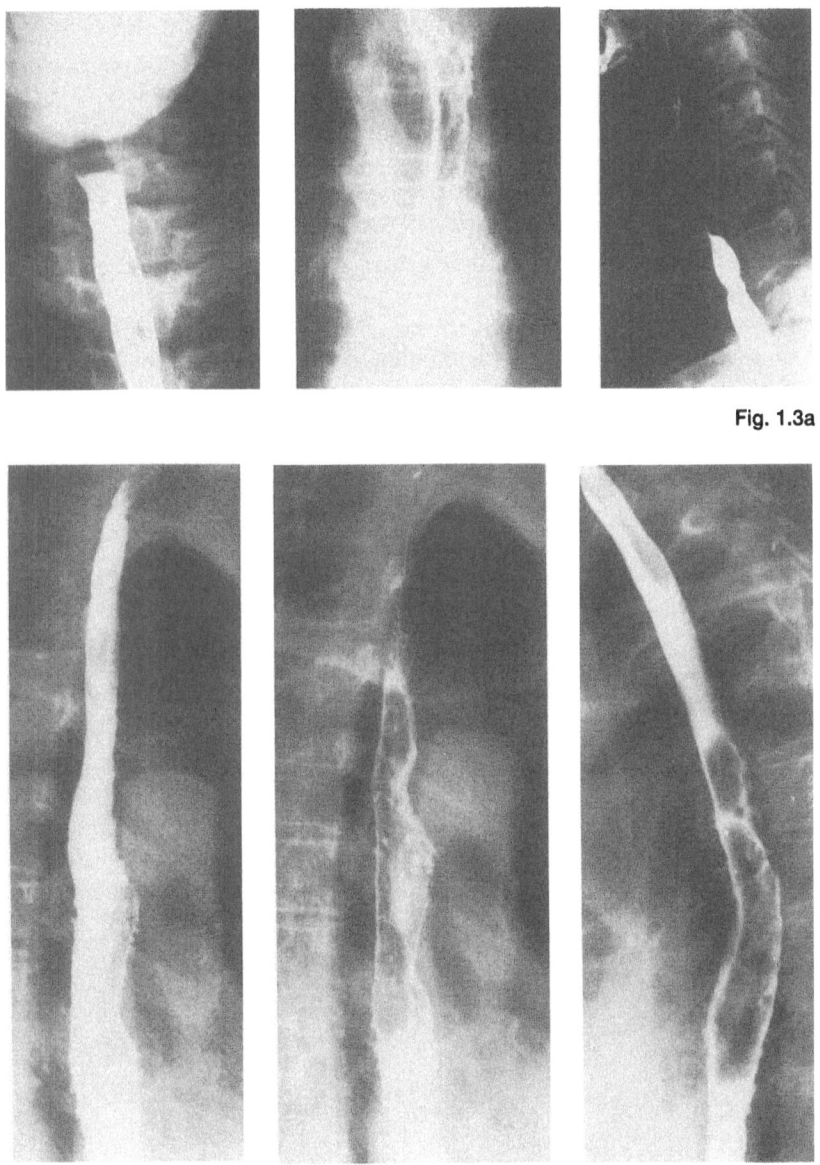

Fig. 1.3a

Fig. 1.3b

Oesophageal Crohn's Disease

Case Report

Barium Swallow: Multiple small pools of barium are seen projecting beyond the outlines of the mid and upper oesophagus which have the appearance of collar-stud ulceration. There are also several linear ulcers in the mid oesophagus. A localised concentric narrowing of the oesophagus at the level of C7 is present, suggestive of local infiltration and mucosal thickening at this site, or possibly an encircling mucosal web.

The appearances of the barium pools are not typical of intramural diverticulosis, and are more those of ulceration.

Differential Diagnosis

1. Oesophageal Crohn's disease; further GI-tract studies would be helpful for further assessment.
2. The changes may be secondary to herpes simplex or cytomegalovirus infection, especially if the patient is immune compromised.
3. Severe candidiasis.

Clinical Data

Neonate with a heart murmur.

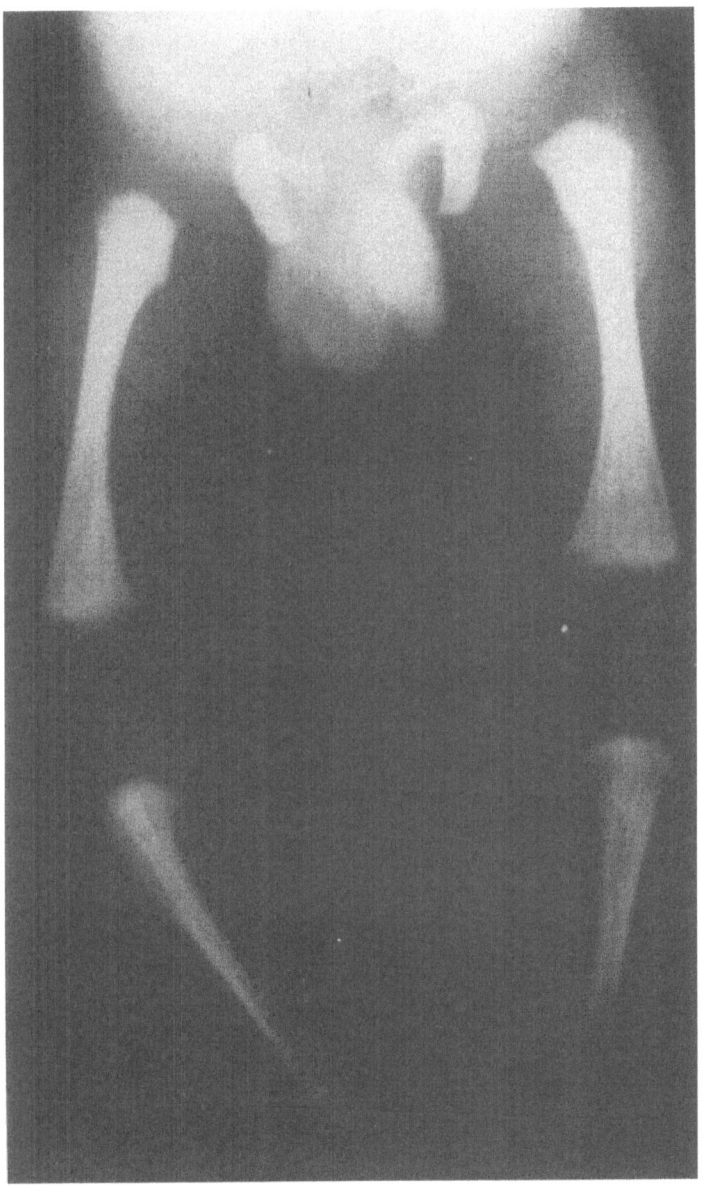

Fig. 1.4

Rubella Osteopathy

Case Report

Lower Limbs: The architecture of the diaphysis and metaphysis of the distal femora and proximal tibiae is disorganised, with longitudinal streaks of lucency and sclerosis. The metaphyses are frayed and irregular.

Diagnosis

The appearances are those of congenital rubella osteopathy, with a classical "celery stick" appearance.

Clinical Data

A 4-week-old girl presenting with vomiting.

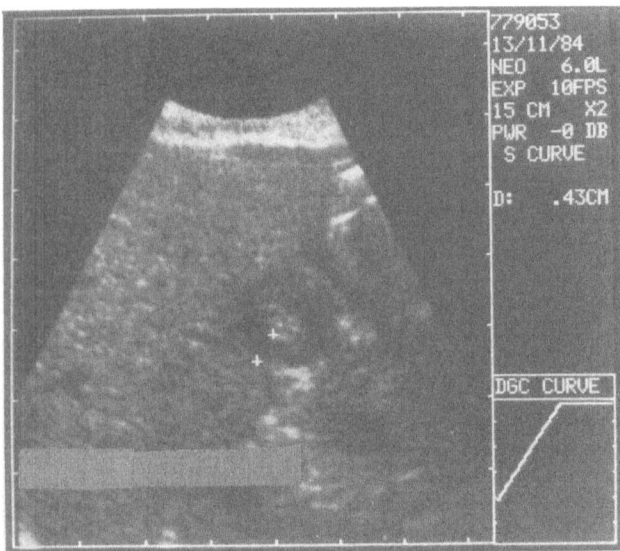

Fig. 1.5a

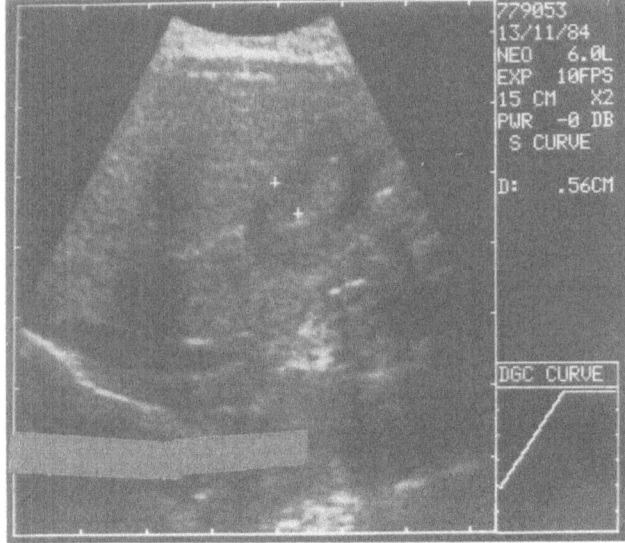

Fig. 1.5b

Pyloric Stenosis

Case Report

Ultrasound of Upper Abdomen: In the transverse view a circular mass is visible which has an echo-poor outer ring surrounding a more reflective inner core. The outer ring has a thickness of 4.3 mm. The longitudinal view shows the mass to be oval, with a similar echo-poor outer wall, 5.6 mm in thickness, with a reflective centre.

The mass lies inferior to the liver.

Diagnosis

The appearances are those of pyloric stenosis with a hypertrophied muscular wall and an elongated pyloric canal.

Clinical Data

Patient, aged 40 years, presenting with generalised aches and pains.

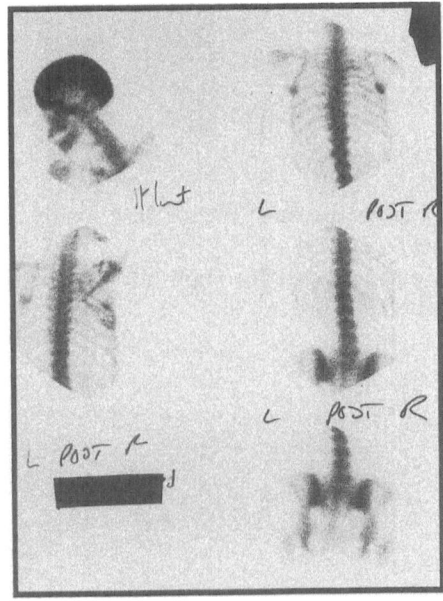

Fig. 1.6a

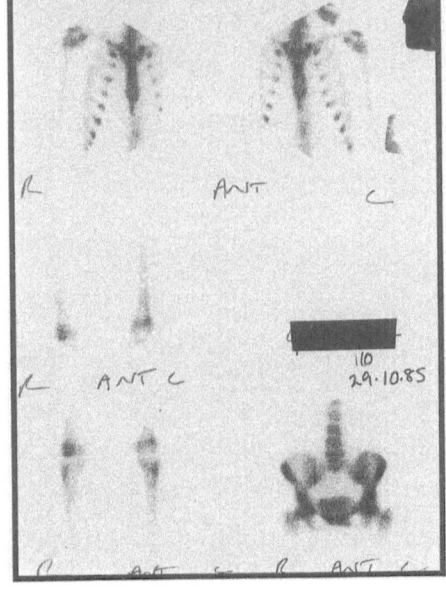

Fig. 1.6b

Osteomalacia

Case Report

Bone Scan: There is generalised increased tracer uptake throughout the skeleton but most marked in the calvarium and the anterior ends of the ribs. The kidneys are only very faintly visualised, but tracer is visible in the bladder.

There is a localised area of increased uptake over the right scapular tip, which, in the arm-raised view, is seen to lie in a rib.

The appearances are those of a "superscan".

Differential Diagnosis

1. Osteomalacia with a pseudofracture in a rib is the most likely cause for these appearances.
2. Metastases are possible, but the increased uptake in the skull relative to the axial skeleton is more typical of a metabolic disorder.

Clinical Data

A 56-year-old female with loin pain.

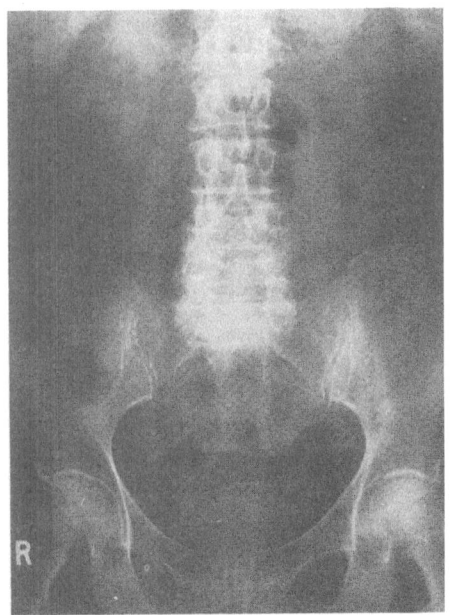
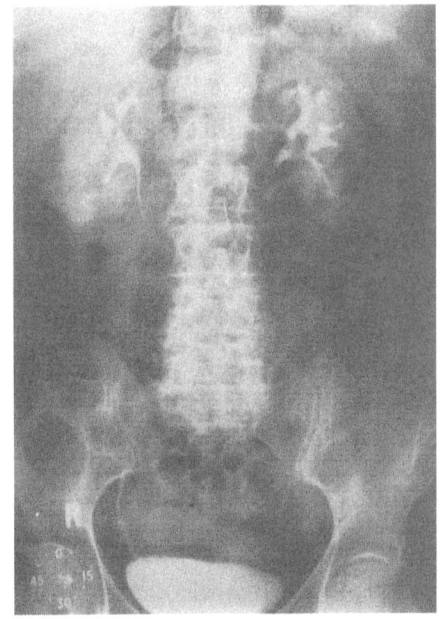

Fig. 1.7a

Fig. 1.7b

Papillary Necrosis Due to Analgesic Abuse

Case Report

IVU Control: Dense calcifications are projected over both renal outlines. The calcifications are more marked on the right and are largely triangular in outline, suggesting calcified papillae. There is severe degenerative disease of the lower lumbar spine.

Full Length Film: Both kidneys are decreased in size and have irregular, undulating outlines. All the calices are very abnormal and are mostly clubbed with numerous cavities filling from them. There is a suggestion of a round intracaliceal mass in the right mid-pole calix.

Differential Diagnosis

The appearances are those of papillary necrosis.

1. Analgesic abuse is a likely cause for the above appearances, particularly in the presence of significant lumbar osteoarthritis.
2. Diabetes and infection.
3. Alcohol-associated papillary necrosis.
4. Sickle-cell anaemia is very unlikely in a middle-aged adult, especially in the absence of typical vertebral changes.
5. Tuberculosis alone could conceivably cause all these changes.

Clinical Data

This 3-year-old child had fallen 20 feet out of a tree.

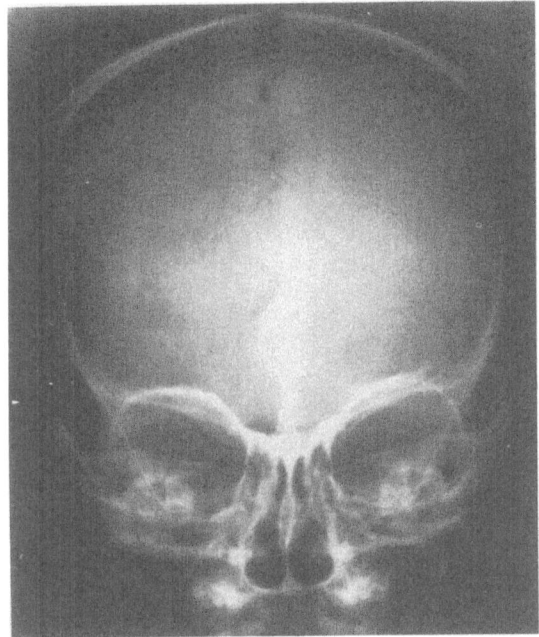

Fig. 1.8a

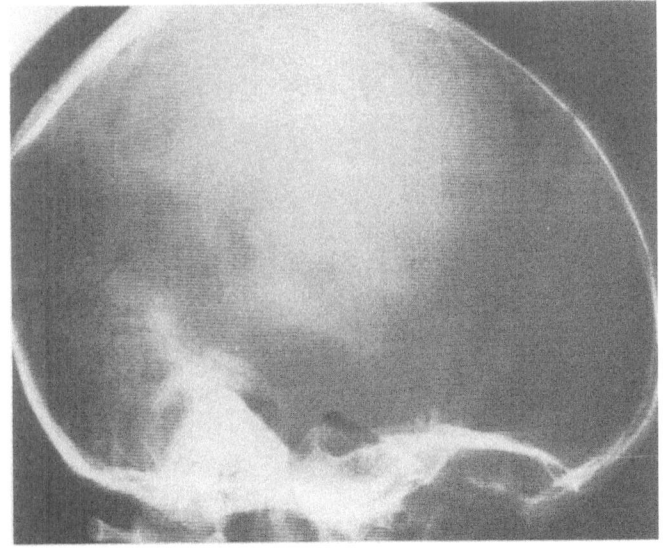

Fig. 1.8b

Skull Fracture With Intracranial Air

Case Report

Skull AP and Lateral: There is a crescent-shaped lucent area lying above the frontal sinuses which is due to intracranial air. There is further intracranial air in the suprasellar region, within the interhemispheric fissure and just above the tentorium. No fracture can be seen although the sphenoid sinus is opaque.

Diagnosis

Cranial fracture, possibly extending into sinuses or skull base. Appropriate plain views or CT of these areas may show the actual fractures.

Supplementary Film

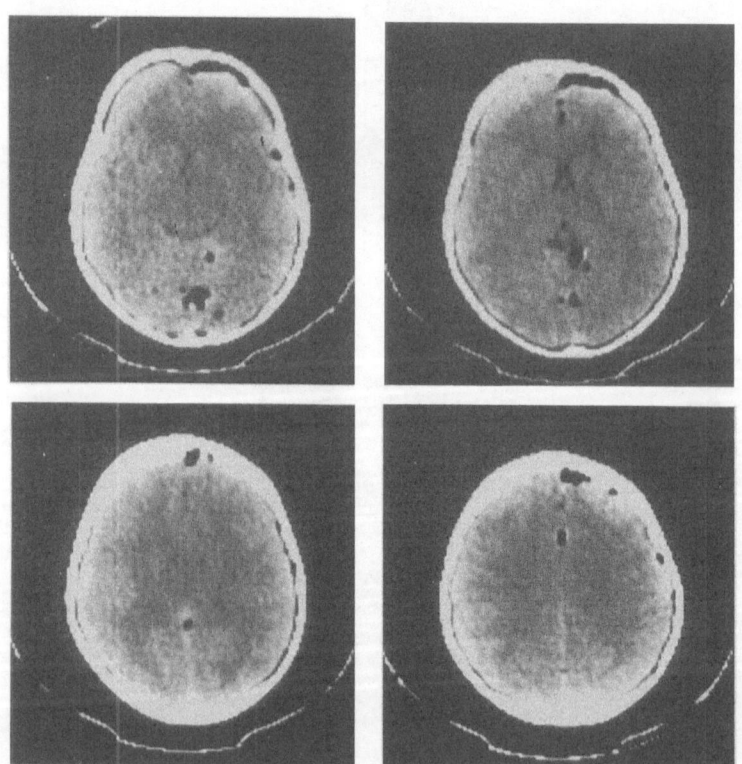

Fig. 1.8c

Head CT: *This confirms the intracranial air but does not show the fractures.*

18

Examination 2

Clinical Data

This 74-year-old man presented with abdominal swelling.

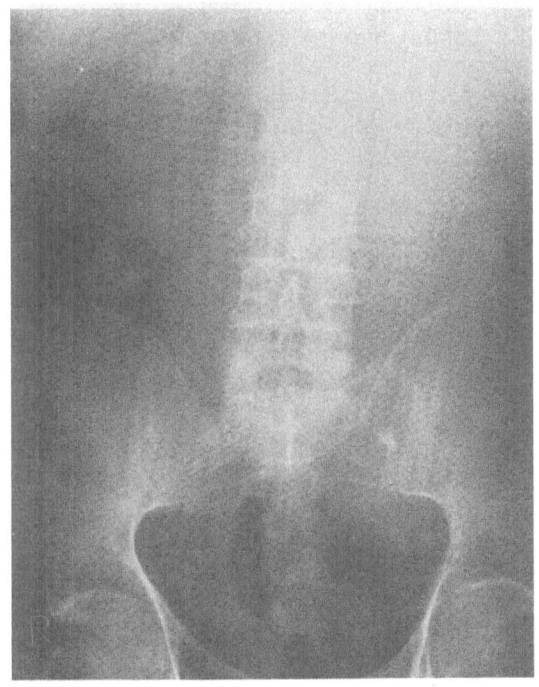

Fig. 2.1a

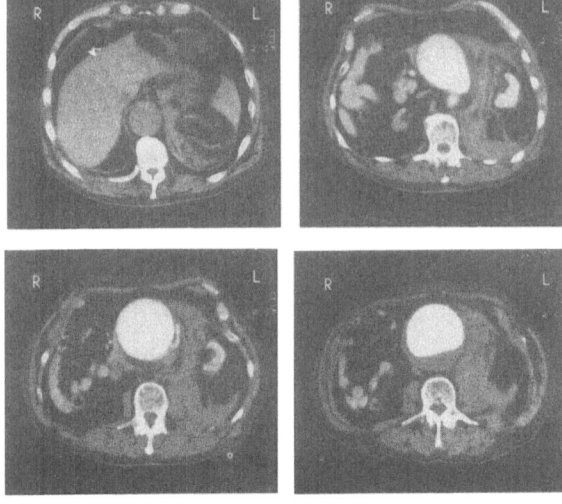

Fig. 2.1b

Leaking Abdominal Aortic Aneurysm

Case Report

Plain Abdomen: The left psoas outline is obscured and there is a large soft-tissue mass in the left side of the abdomen, displacing the bowel gas laterally.

Abdominal CT: Three of the images show contrast enhancement. There is a large abdominal aortic aneurysm extending to the level of the kidneys, associated with a soft-tissue mass extending into the left anterior and posterior pararenal spaces. The fat in this area is abnormally dense and streaky. The left kidney is reduced in size and is displaced anteriorly and laterally but it does excrete contrast, whereas the smaller right kidney shows no excretion at all.

Diagnosis

The appearances are of an abdominal aortic aneurysm with bleeding into the left retroperitoneum.

Clinical Data

This 18-year-old female presented with acute shortness of breath.

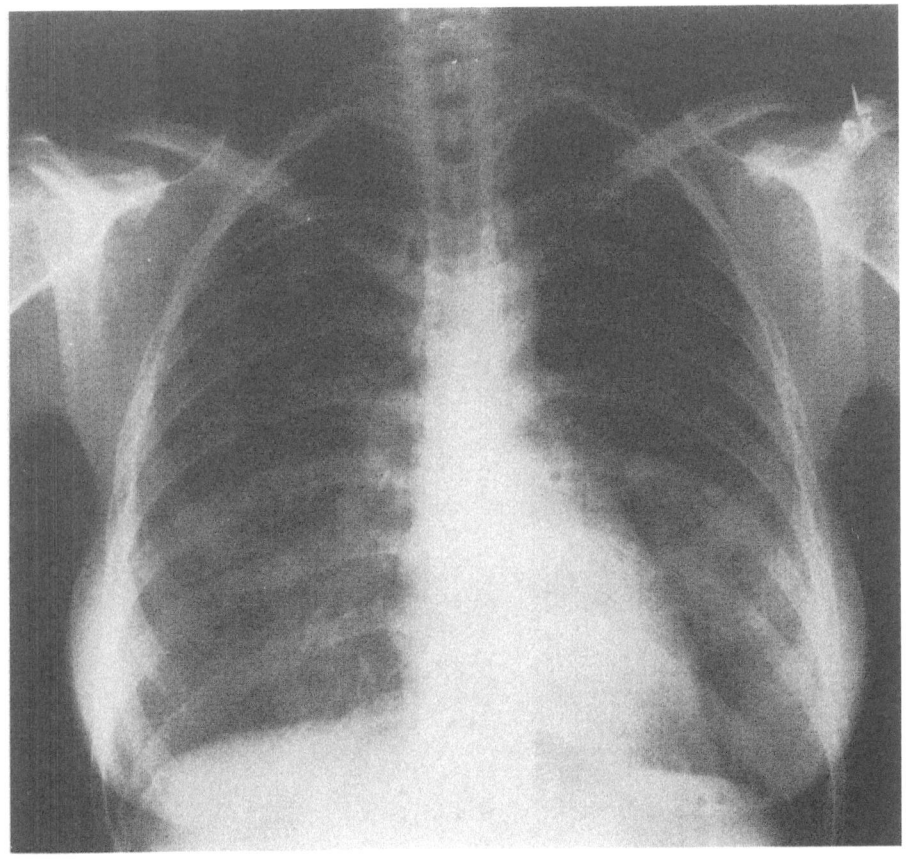

Fig. 2.2

Chickenpox Pneumonia

Case Report

PA Chest: There is widespread, symmetrical alveolar type density, predominantly in the mid and lower zones, with a poorly defined air bronchogram. The heart size is normal and the upper lobe vessels are normal in size.

Differential Diagnosis

The differential diagnosis for this appearance is large but the acute history reduces the possibilities to those below.

1. Infective causes: Viral, mycoplasma, or possibly bacterial. It is important to know if the patient is immune-suppressed, so that causes such as pneumocystis can be considered.
2. Non-cardiogenic pulmonary oedema due to drugs, noxious gas inhalation, fluid overload, etc.
3. Haemorrhage due to bleeding disorders or Goodpasture's syndrome.
4. Fat embolism and adult respiratory distress syndrome.

Clinical Data

A 43-year-old female presenting with anaemia.

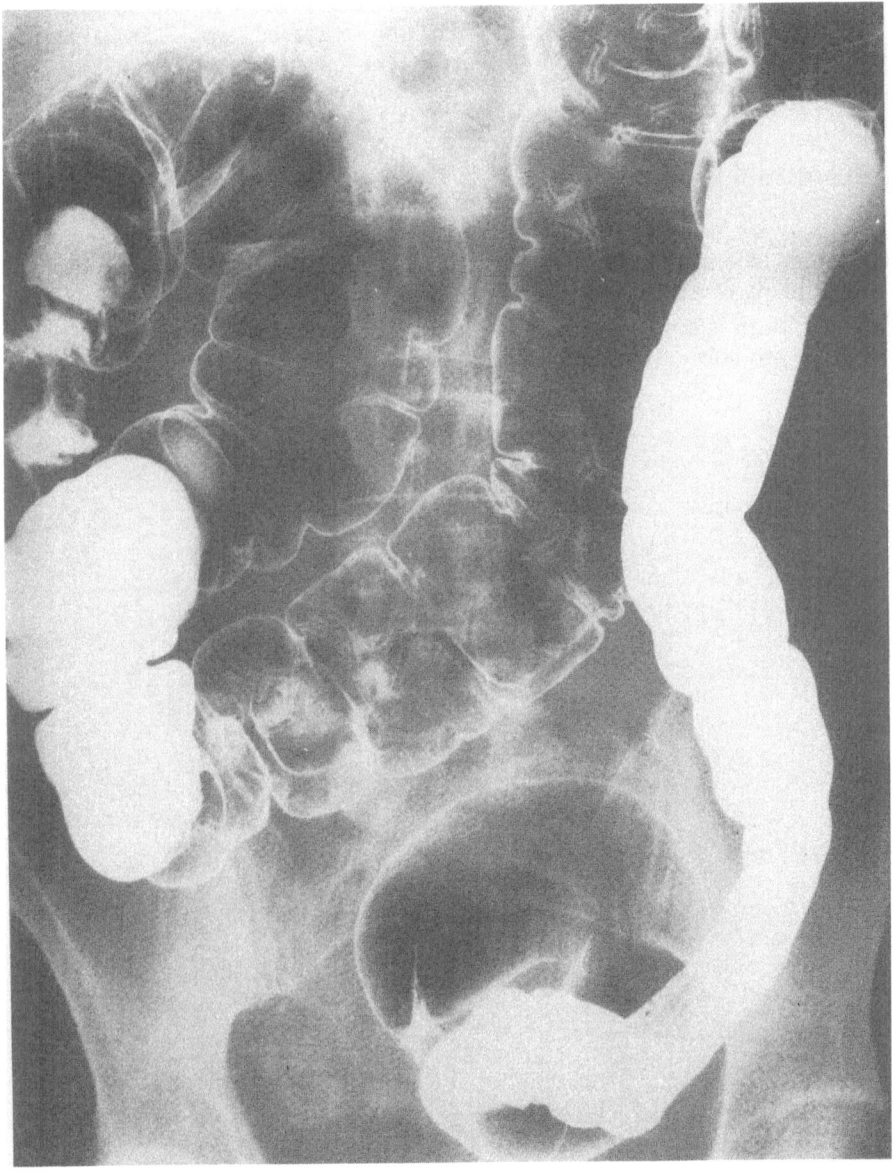

Fig. 2.3

Chronic Ulcerative Colitis with Complicating Carcinoma

Case Report

Double-contrast barium enema: There are widespread linear filling defects seen largely in the transverse colon which represent "filiform polyps". The haustral pattern is normal, and no abnormality is seen in the rectum. The ascending colon is distorted into a 4-cm long strictured "apple core" segment with destroyed mucosa within this area. The caecum is not visualised. The sacroiliac joints appear normal.

Differential Diagnosis

1. Chronic ulcerative colitis complicated by a carcinoma in the right colon is the most likely cause for these appearances.
2. The pattern is not typical for familial polyposis coli, where the adenomatous polyps are normally rounded and more numerous.

Clinical Data

A 25-year-old female with pain in the left hip.

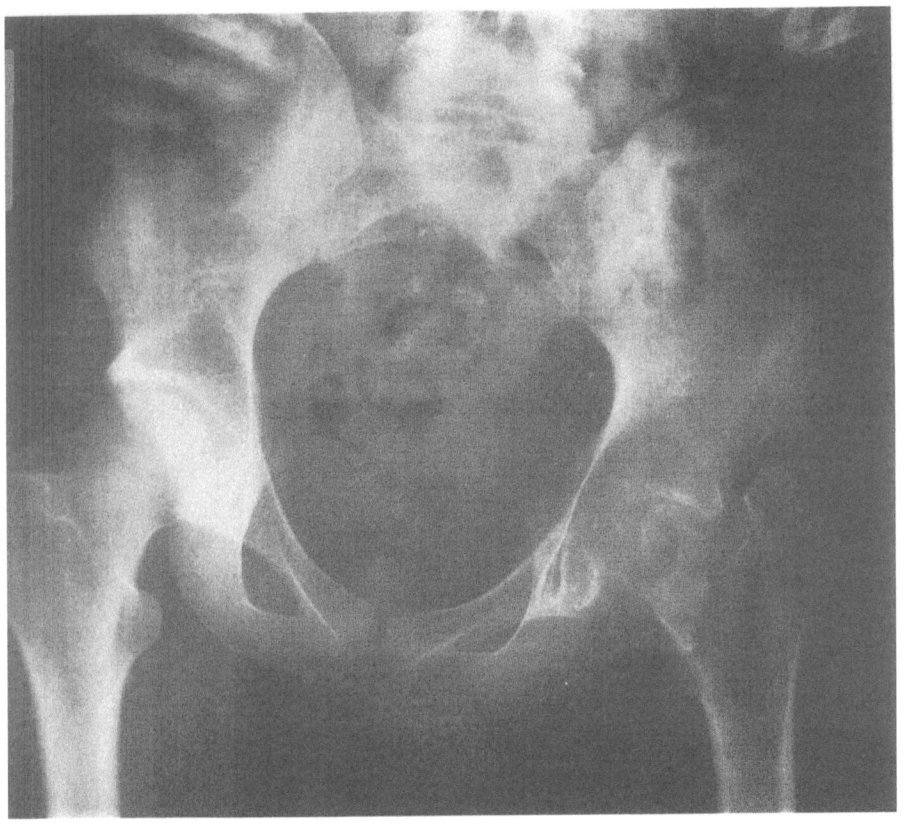

Fig. 2.4

27

Giant-Cell Tumour With Pathological Fracture

Case Report

Pelvis: The left upper femur shows a large area of poorly defined lucency extending from the upper shaft to the greater trochanter and into the femoral head. There is an intertrochanteric fracture which is presumably pathological. The lucent area has a wide zone of transition and there is no associated bony expansion, periosteal reaction, or soft-tissue mass. These appearances suggest a relatively aggressive lesion. The left femoral shaft below the lesion and the left ilium show marked osteopenia which may represent disuse osteoporosis.

Differential Diagnosis

1. Giant-cell tumour is likely. Although this lesion does not reach the joint margin it does extend up to an apophyseal margin (greater trochanter).
2. Primary bone tumour such as a lytic osteosarcoma.
3. Metastasis.
4. Osteomyelitis.
5. Possibly an aggressive form of histiocytosis X.

Clinical Data

A 48-year-old male who suffered from night sweats.

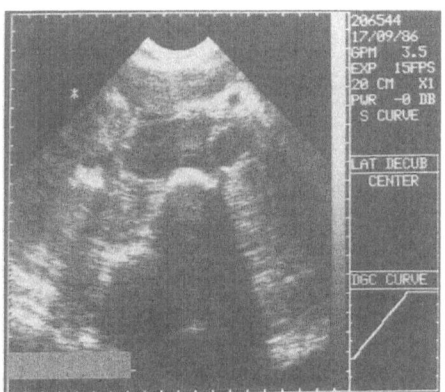

Fig. 2.5a

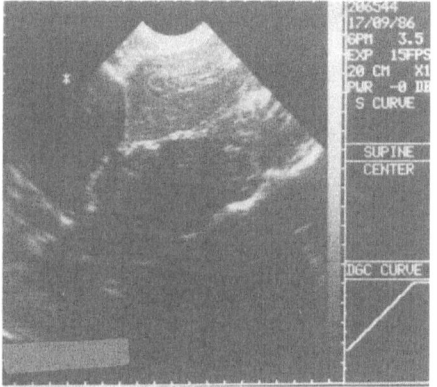

Fig. 2.5b

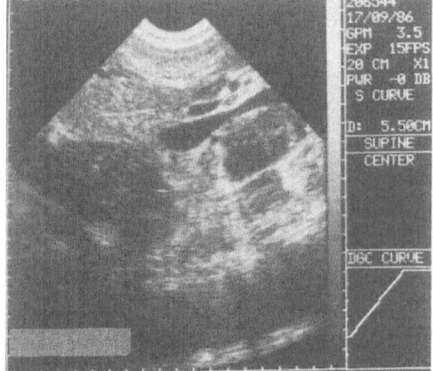

Fig. 2.5c

Para-aortic Nodes Due to Hodgkin's Disease

Case Report

Abdominal Ultrasound: There are large hypoechoic masses anterior and posterior to the IVC and around the aorta. These masses contain some echoes and are therefore of soft-tissue composition. There is extension up to the diaphragm in the retrocrural region.

These masses represent para-aortic and retrocrural lymphadenopathy. CT would help define the extent of the abnormality.

Differential Diagnosis

1. Lymphoma is the most likely cause for such marked, widespread lymphadeno-pathy.
2. Metastases are less likely to cause such adenopathy.

Supplementary Film

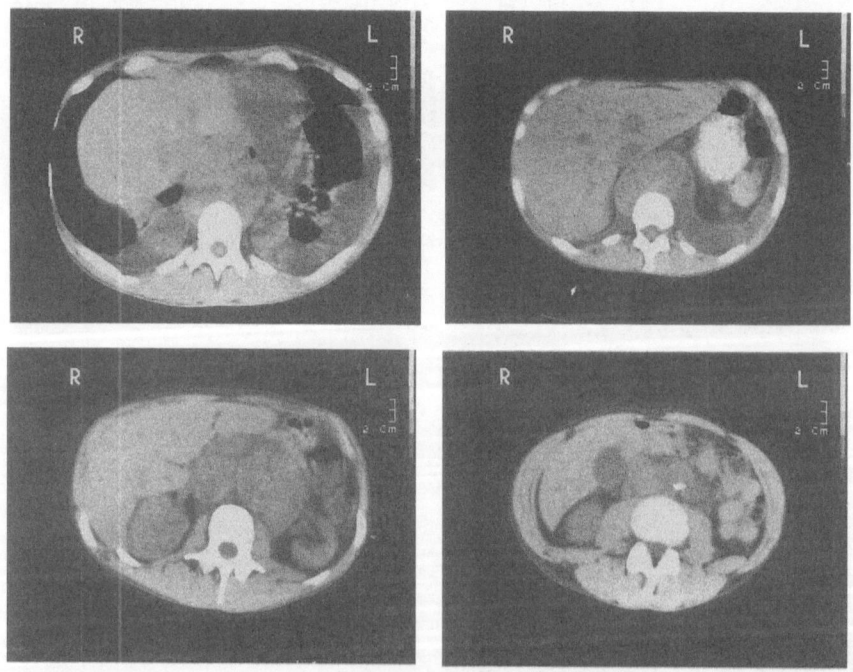

Fig. 2.5d

Abdominal CT: *Bilateral pleural effusions are present and the massive para-aortic and retrocrural lymphadenopathy is confirmed.*

Clinical Data

Male, 25 years old, presenting with painful feet and back.

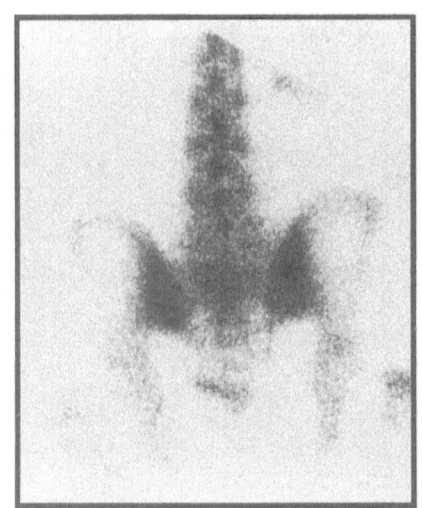

Fig. 2.6a

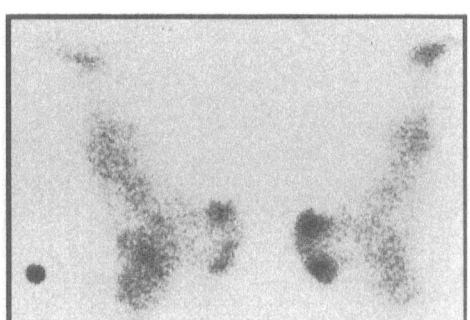

Fig. 2.6b

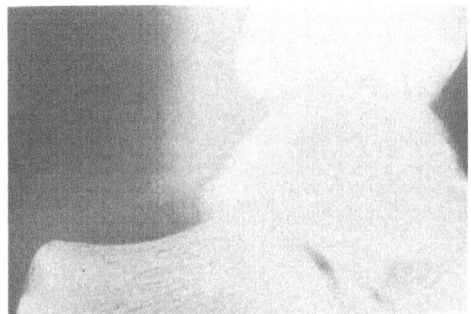

Fig. 2.6c

Psoriasis

Case Report

Bone Scan, Pelvis and Feet: There is increased tracer uptake in the left sacroiliac joint, and also to a lesser extent in the right. The feet show markedly abnormal tracer uptake in the left calcaneum at the sites of insertion of the Achilles tendon and the plantar fascia. Similar but less marked changes are seen in the right calcaneum.

Lateral Soft-Tissue of Left Ankle: There is a small erosion at the insertion of the Achilles tendon, and soft-tissue swelling is seen associated with the distal part of the tendon, causing loss of the normal lucent space adjacent to the postero-superior margin of the calcaneum.

Differential Diagnosis

1. Seronegative arthritis such as psoriasis, Reiter's disease, or ankylosing spondylitis are most likely with an asymmetrical distribution.
2. Rheumatoid arthritis is possible but is usually more symmetrical and presents with small-joint involvement.
3. Stress-related injuries are also possible.

Clinical Data

Neonate with vomiting.

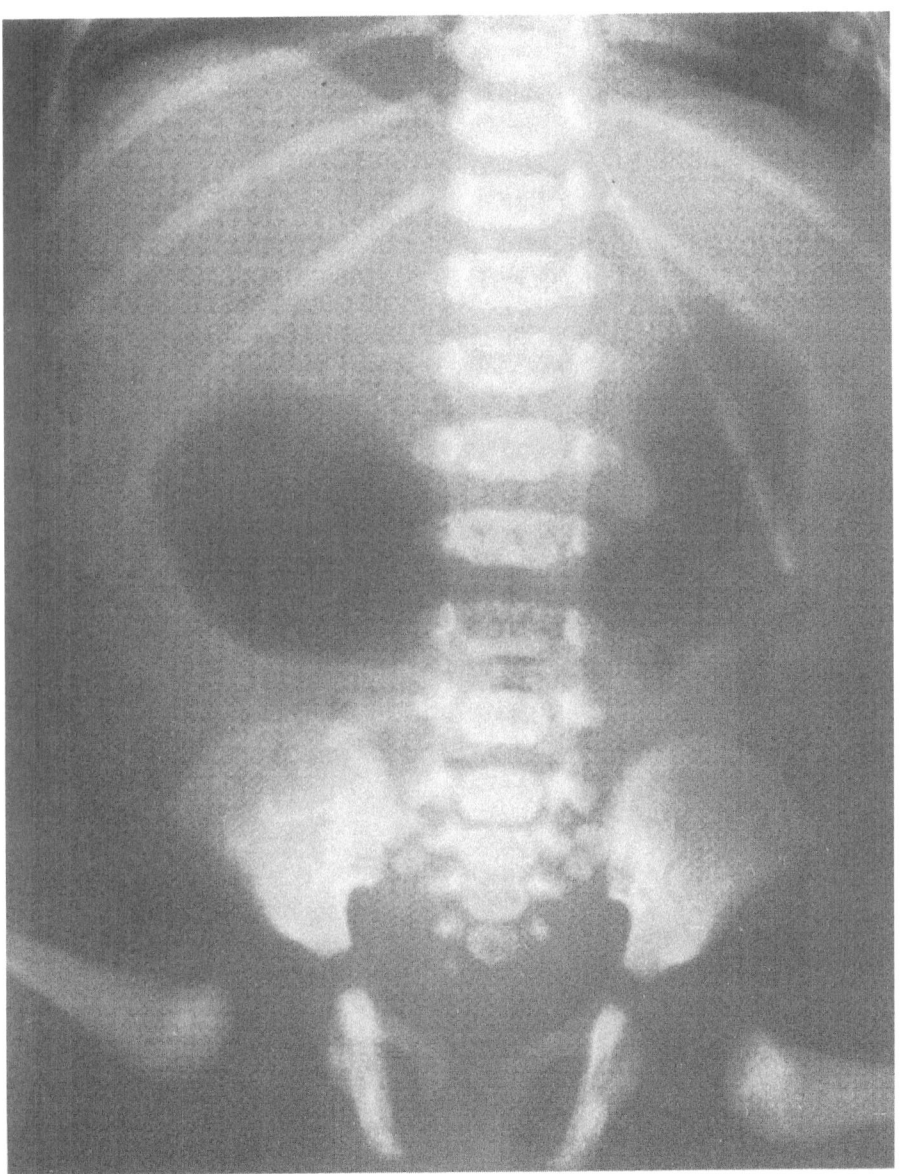

Fig. 2.7

Duodenal Atresia with Ileal Atresia and Meconium Perforation

Case Report

Plain Abdomen: Two dilated gas-containing structures are visible in the upper abdomen, producing a "double bubble" appearance. A very small amount of gas is seen distally. Six lumbar vertebrae are present.

Faint calcification is seen in the right upper quadrant, which is very suggestive of an in-utero meconium perforation.

Differential Diagnosis

1. A double bubble sign is usually associated with duodenal atresia, although other causes of congenital duodenal narrowing such as duodenal webs, bands and (much less likely) annular pancreas, should also be considered.
2. A meconium perforation is unlikely to be due to duodenal atresia and is probably due to either an associated distal small-bowel atresia or meconium ileus.

Clinical Data

A 16-year-old boy with a frontal swelling.

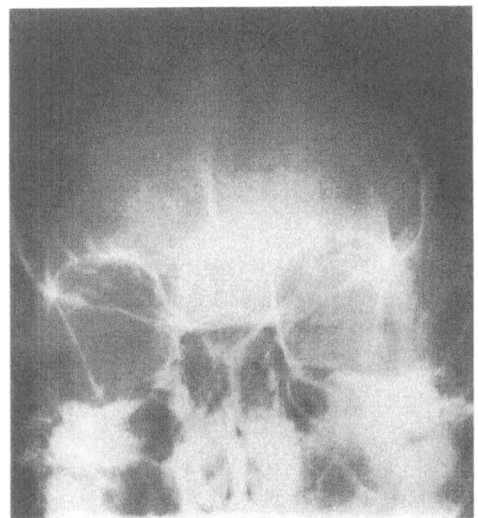

Fig. 2.8a

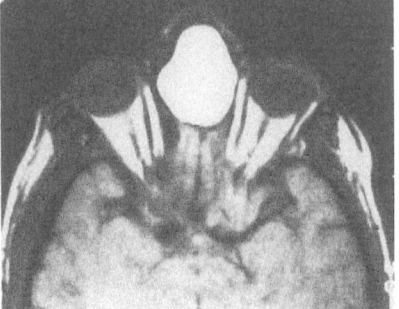

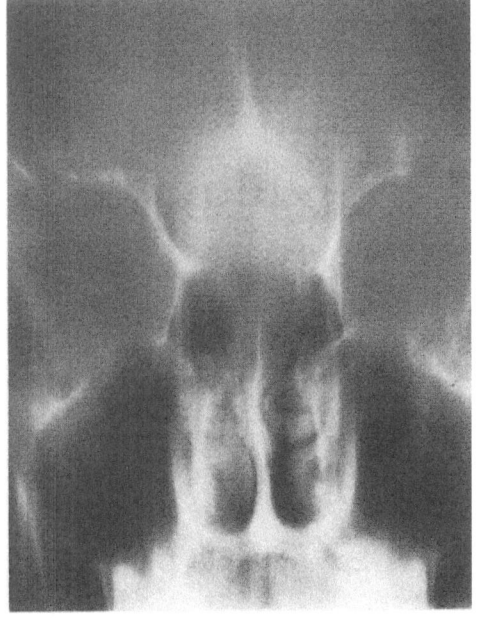

Fig. 2.8b

Fig. 2.8c

Frontal Dermoid

Case Report

Sinus View and Plain Frontal Tomogram: There is marked hypertelorism and a well-defined lytic lesion with sclerotic margins in the frontal bone lying between the orbits. The upper part of the nasal septum is eroded, with a suggestion of a soft-tissue mass lying within the abnormal area.

MRI of Face: There is an area of very high signal intensity between the orbits and extending upwards from the upper part of the nasal cavities; this is of the same signal strength as the intraorbital fat.

Differential Diagnosis

1. Frontal dermoid and lipoma are both equally likely.
2. Encephalocele is likely from the plain films, but the high fat content on MRI is against this.

Examination 3

Clinical Data

Pre-employment chest film of a 30-year-old male.

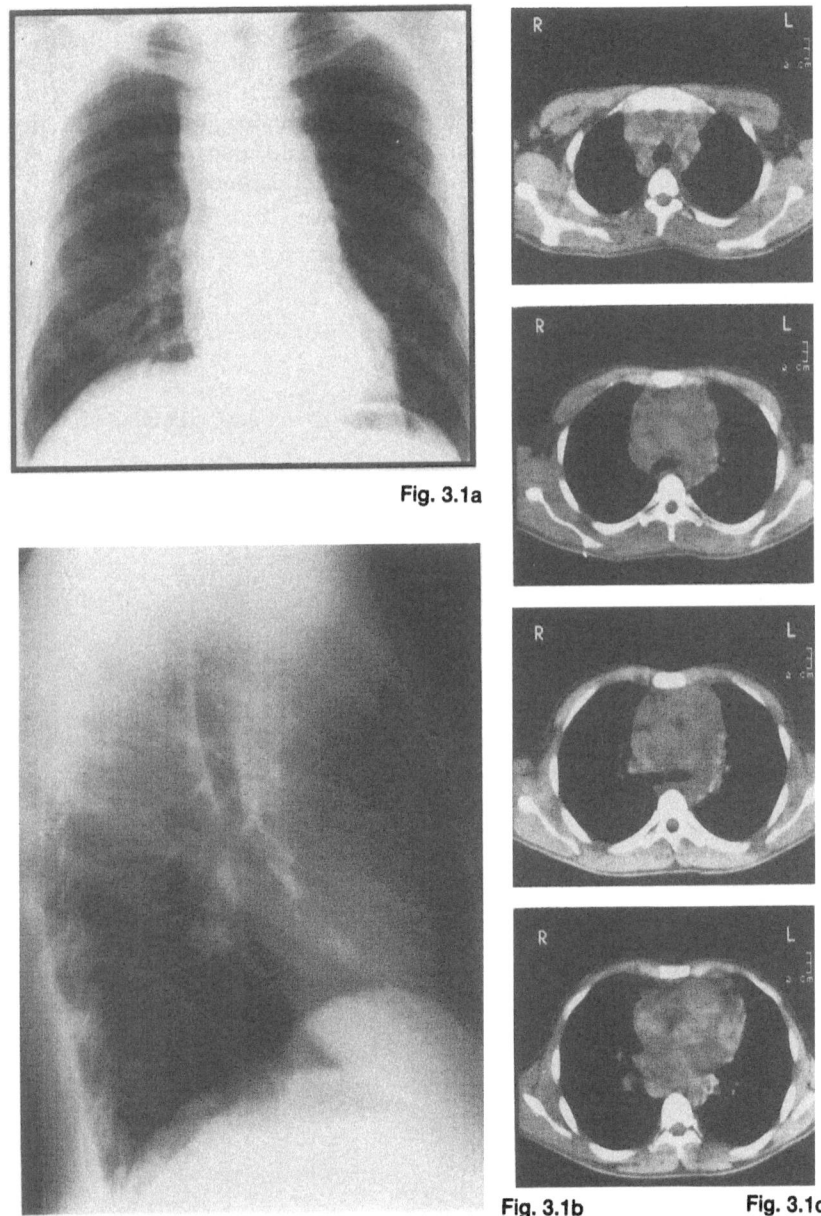

Fig. 3.1a

Fig. 3.1b   Fig. 3.1c

Hodgkin's Disease

Case Report

PA Chest: The superior mediastinum is markedly widened, with gross thickening of the paratracheal area and obscuration of the ascending aorta. The mass does not extend above the clavicles, suggesting that it is largely anterior. No rib or lung abnormality is seen.

Lateral Chest: The anterior window is filled in, confirming the anterior position of the mass.

CT Chest (Unenhanced): Numerous enlarged lymph nodes are visible anterior to the aortic arch and in the aortopulmonary, azygos, and subcarinal regions. All the major vessels are displaced posteriorly. The appearances are of massive mediastinal lymphadenopathy.

Differential Diagnosis

1. Lymphoma, particularly Hodgkin's disease.
2. Metastatic lymphadenopathy is possible but is much less likely in this age group.

Clinical Data

Post-football injury in a 16-year-old male.

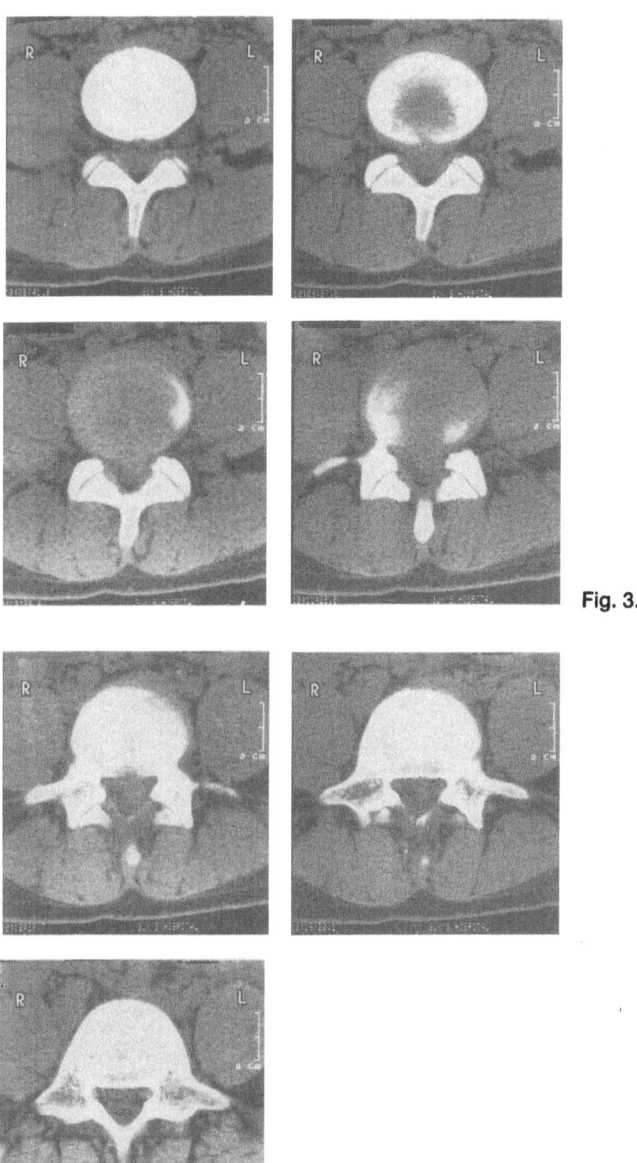

Fig. 3.2a

Fig. 3.2b

Prolapsed L4–L5 Intervertebral Disc

Case Report

CT Lumbar Spine: There is protrusion of soft tissue into the spinal canal at L4–L5 which is continuous with the intervertebral disc at this level. The thecal sac is indented, distorted, and displaced to the right. The normal fat planes surrounding the left L4 nerve root are effaced.

Diagnosis

The appearances are of a prolapsed L4–L5 intervertebral disc.

Clinical Data

Female, 32 years old, with pain in the left shoulder.

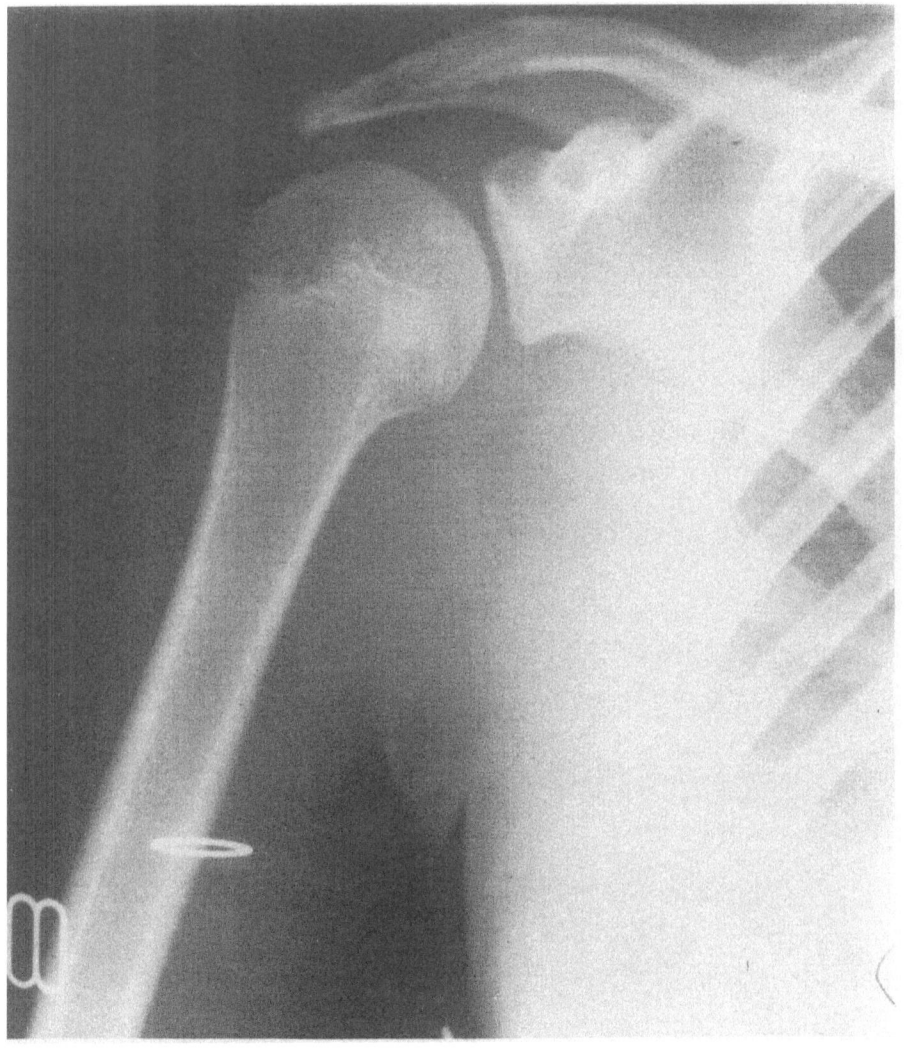

Fig. 3.3a

Posterior Dislocation of the Humerus

Case Report

Left shoulder: There is widening of the distance between the anterior glenoid margin and the medial margin of the humeral head. Also, the normal overlap of the medial part of the humeral head and the glenoid is lost and there is discontinuity of Maloney's line. The humeral head shows slight internal rotation. The glenohumeral orientation is abnormal and a posterior dislocation is likely. An axial film would confirm this.

Supplementary Film

Fig. 3.3b

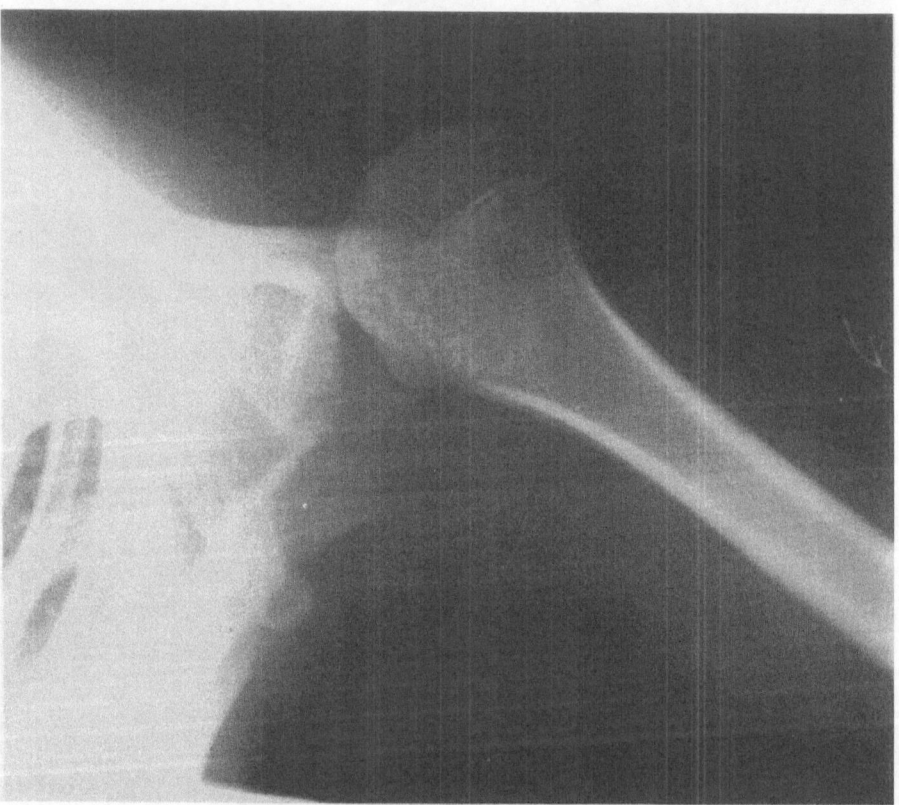

Axial Left Shoulder: *This confirms that there is a posterior dislocation of the humeral head.*

Clinical Data

Female, 30 years old, with right-sided abdominal pain.

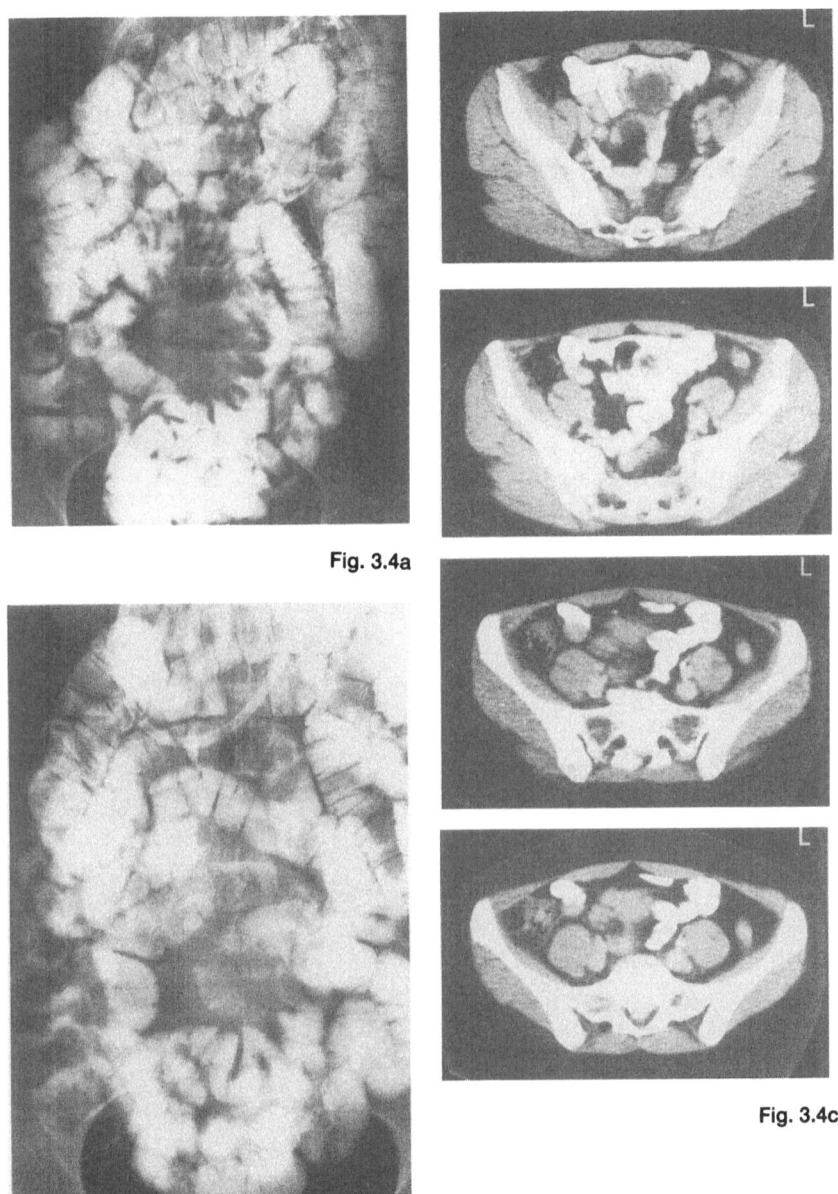

Fig. 3.4a

Fig. 3.4c

Fig. 3.4b

45

Peritoneal Desmoid

Case Report

Small-Bowel Enema: There is a constant filling defect in the centre of the abdomen at the level of L5, displacing bowel loops laterally around it. The mesenteric margins of the loops surrounding this area are indented, stretched, and distorted, particularly on the left, suggesting a localised fibrotic mesenteric reaction.

The diameter and fold thickness of the remaining loops are normal.

CT Abdomen: There is a large central abdominal soft-tissue mass which does not contain contrast, despite good filling of the whole bowel, suggesting that this mass lies outside, but closely related to, the small bowel. No lymphadenopathy is seen.

Differential Diagnosis

1. Carcinoid.
2. Lymphoma.
3. Metastases.
4. Mesenteric inflammatory mass.

All above are approximately equally likely.

5. Peritoneal desmoid: this is possible but is an uncommon cause of these appearances.

Clinical Data

Abdominal mass in a 3-week-old boy.

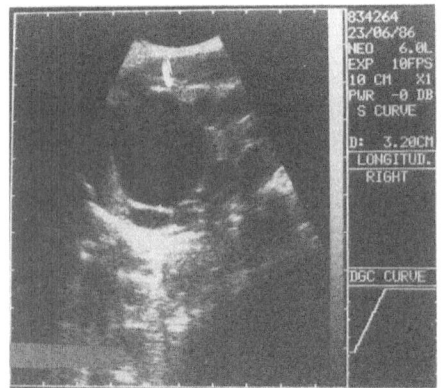

Fig. 3.5a

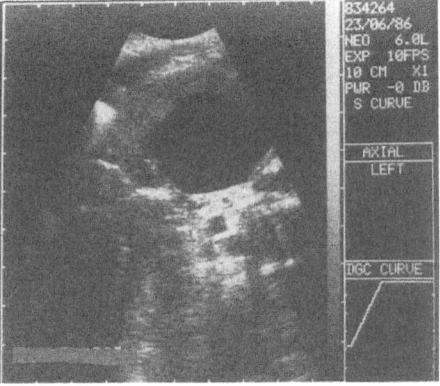

Fig. 3.5b

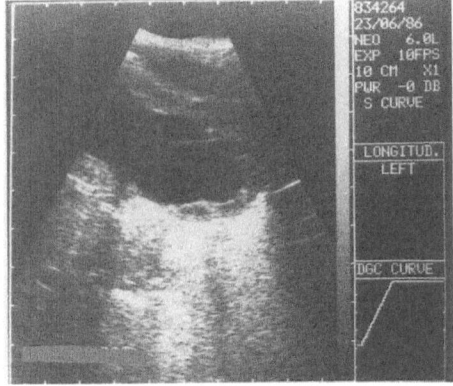

Fig. 3.5c

Fig. 3.5d

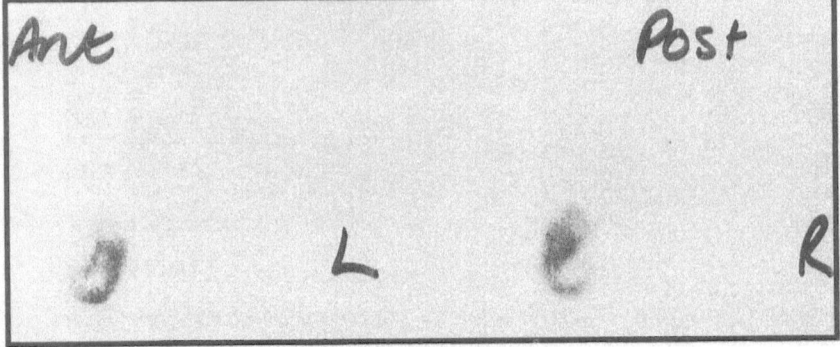

Right Cystic Dysplastic Kidney with Left Pelviureteric Junction Obstruction

Case Report

Ultrasound: The right kidney contains multiple, large, unconnected cystic masses up to 3 cm in diameter. No parenchyma is seen on the right. The left kidney has a massively dilated pelvicaliceal system but retains a reasonable thickness of parenchyma.

DMSA Scan: No right renal function. The left kidney concentrates adequately but contains large photon-deficient areas which correspond to the dilated calices.

Diagnosis

Right cystic dysplastic kidney.
Left pelviureteric junction obstruction.

Supplementary Films

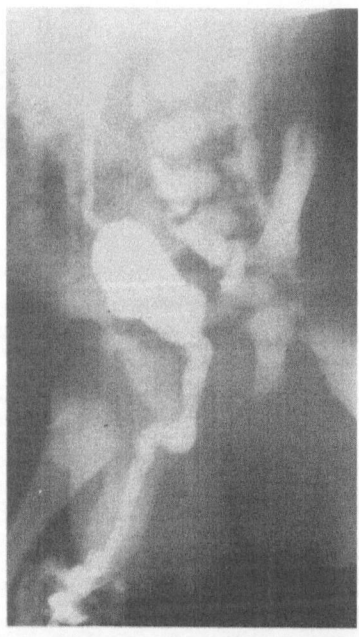 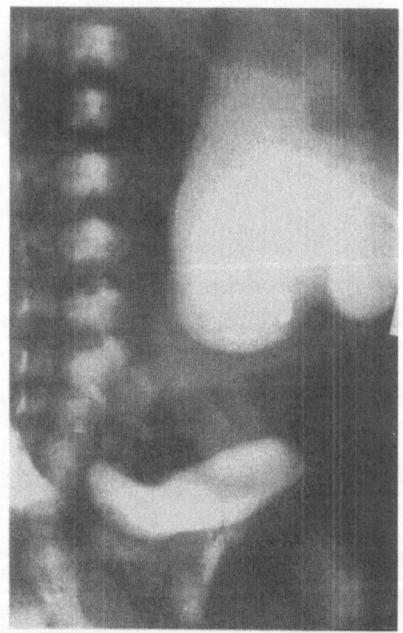

| Fig. 3.5e | Fig. 3.5f |

Micturating Cystogram: *There is bilateral reflux and an atretic right ureter.*
Left Antegrade: *This confirms left pelviureteric obstruction.*

Clinical Data

A 72-year-old male with abdominal distension.

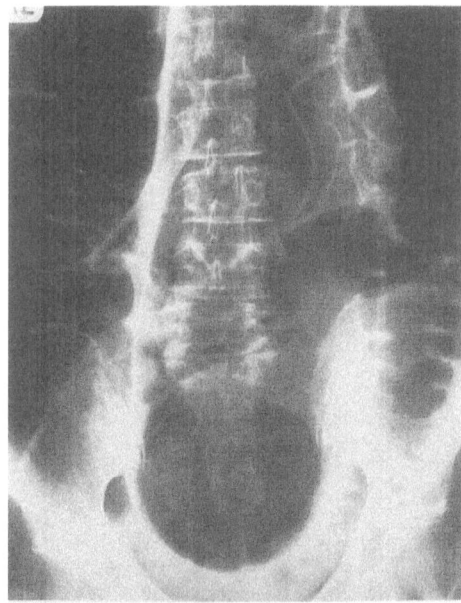

Fig. 3.6a

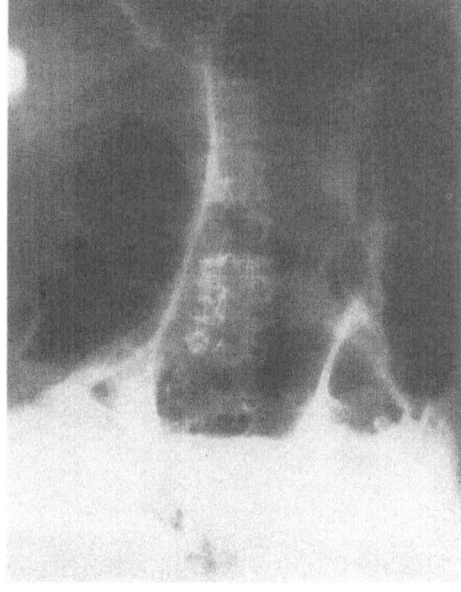

Fig. 3.6b

Sigmoid Volvulus

Case Report

Plain Abdominal Films, Erect and Supine: Massively dilated loops of large bowel are seen lying centrally, and extending down into the pelvis and up into the left upper quadrant. There are also loops of dilated small bowel visible in the right side of the abdomen. The caecum appears in the normal position.

Diagnosis

The appearances are typical of a sigmoid volvulus, and a barium enema may confirm this if the patient is well enough.

Clinical Data

Male, 12 years old, with abdominal pain.

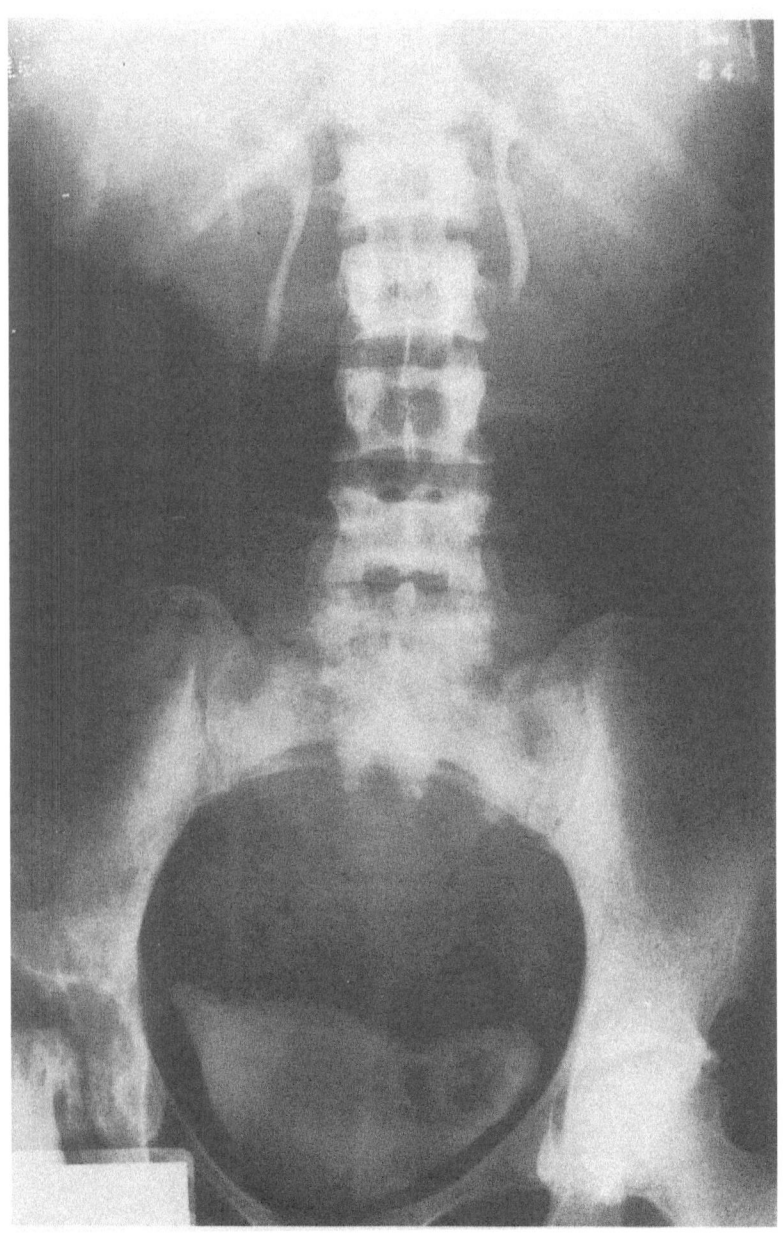

Fig. 3.7

Sickle-Cell Disease

Case Report

IVU (15-Minute Film): No control film is available. There are calcific opacities in the right upper quadrant which are most probably gallstones. The lumbar vertebrae, especially L1 and L2, show central depression of the end-plates.

The right femoral head is irregularly destroyed and sclerotic.

The calices of both kidneys are poorly seen but appear deformed and clubbed.

Diagnosis

The appearances are those of sickle-cell disease, causing central depression of vertebral end-plates, avascular necrosis of the right femoral head, and gallstones. The caliceal appearances probably represent papillary necrosis, but further films are needed to confirm this.

Clinical Data

A 5-year-old girl with a painful torticollis.

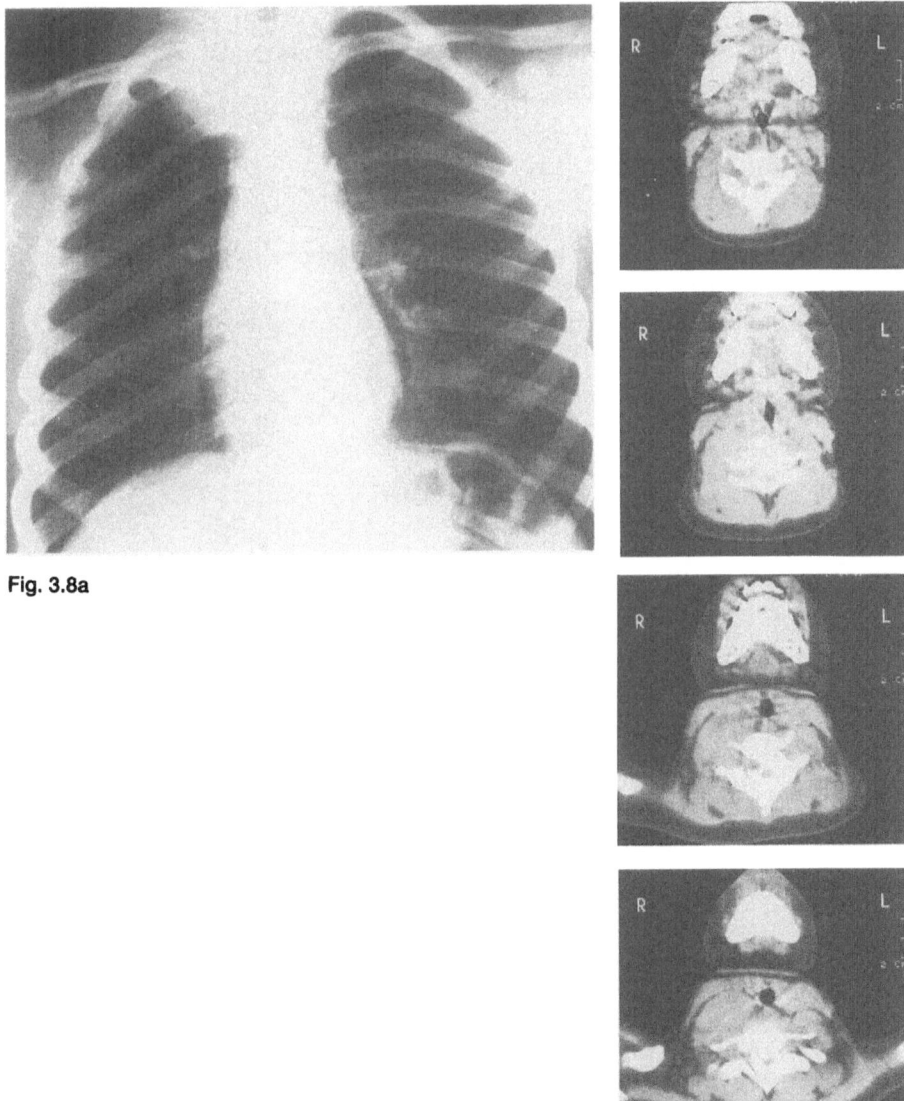

Fig. 3.8a

Fig. 3.8b

Neuroblastoma

Case Report

PA Chest: There is a right superior mediastinal mass, which lies posteriorly since it extends above the clavicle. There is no visible associated rib or vertebral involvement.

CT Cervical Spine (With Intrathecal Contrast): There is a large soft-tissue mass lying anterior to the right side of the visible vertebral bodies, with extension of soft tissue into the spinal canal, causing widening and erosion of the right intervertebral foramen and indenting and displacing the thecal sac posteriorly.

The appearances are of a cervical extradural mass and a posterosuperior mediastinal mass.

Differential Diagnosis

1. Neuroblastoma is the most likely cause for the above in this age group.
2. Neurofibromatosis is less likely at this age.
3. Paragangliomas should also be considered.

Examination 4

Clinical Data

Female, 45 years old, with mild shortness of breath.

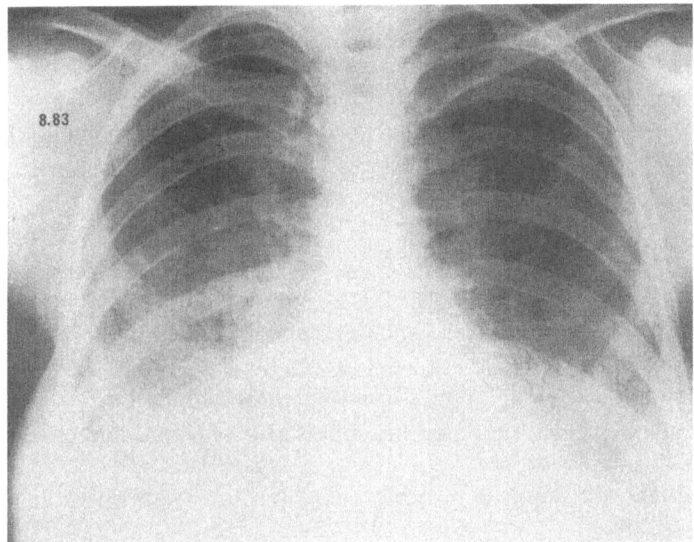

Fig. 4.1a

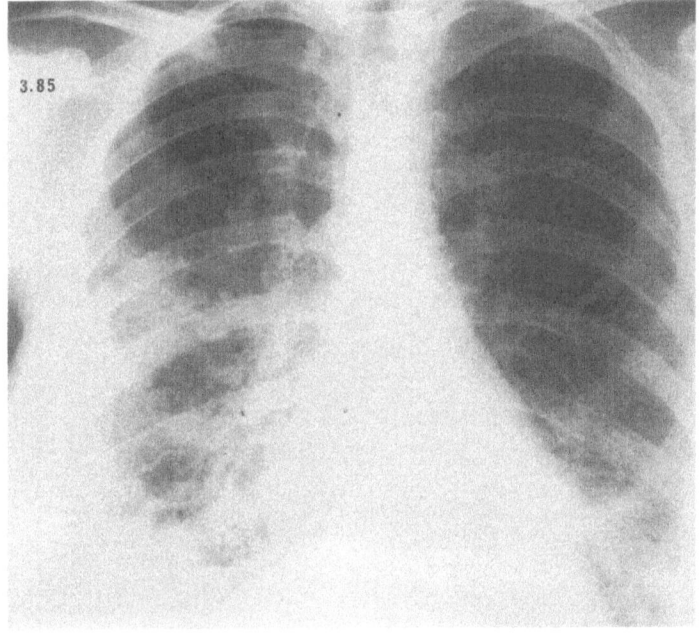

Fig. 4.1b

Alveolar Proteinosis

Case Report

PA Chest: There is widespread, poorly defined airspace density with an air bronchogram, most marked in the mid and lower zones. Little change in appearance is seen between the two films, which were obtained 18 months apart. The heart is normal in size and there are no signs of raised pulmonary venous pressure, although the diffuse alveolar type density is very suggestive of pulmonary oedema.

The chronic nature of the underlying disease process and the lack of severe symptoms in the presence of substantial radiological changes limit the differential diagnosis.

Differential Diagnosis

1. Alveolar proteinosis is a rare disease which presents in this fashion.
2. Alveolar-cell carcinoma is possible but would probably have progressed more over this time course. The same applies to an indolent lymphoma.
3. Sarcoid with an alveolar distribution occurs, but is also very rare and tends to involve the mid and upper zones.
4. Recurrent low-grade aspiration is possible and a barium swallow would be useful in assessing this.

Clinical Data

Male, 45 years old, with loss of sight in the left eye.

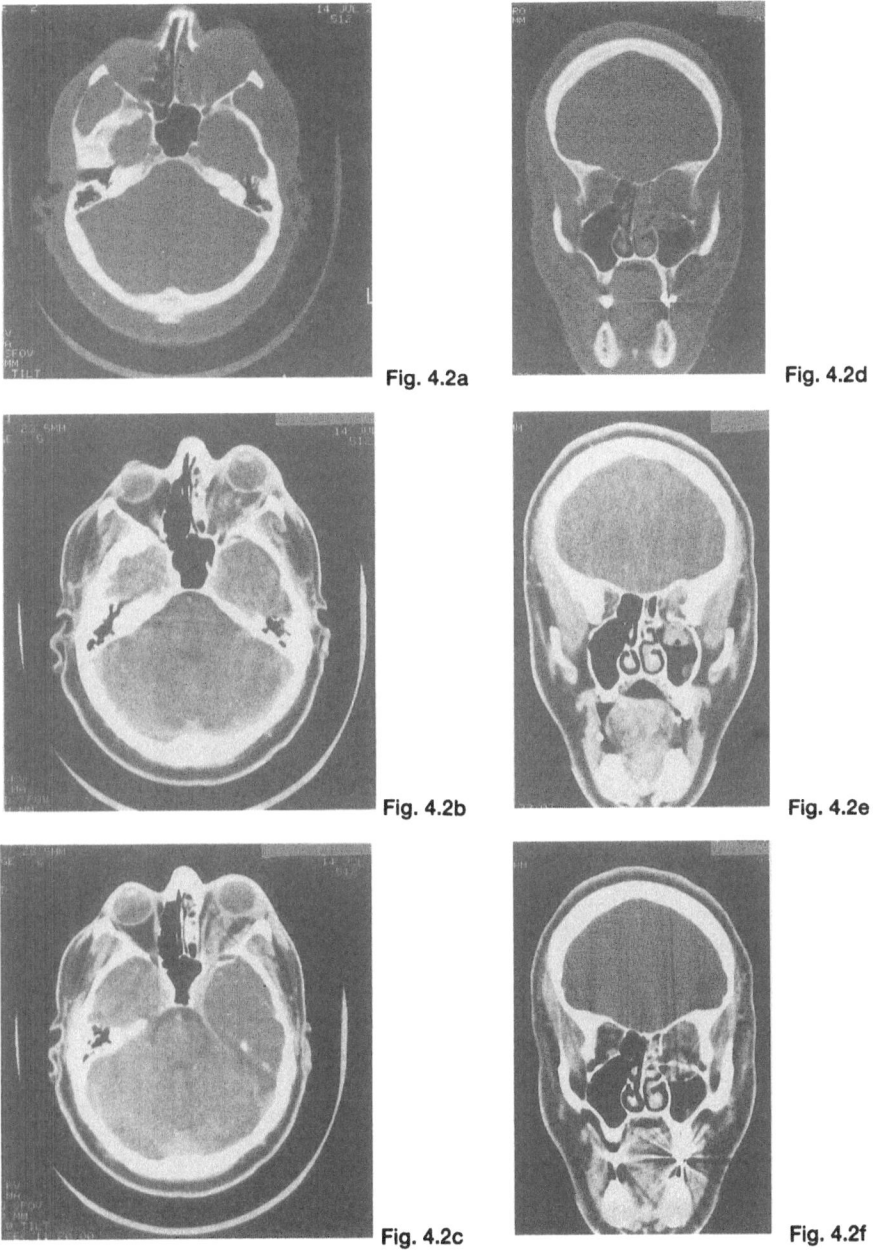

Fig. 4.2a

Fig. 4.2d

Fig. 4.2b

Fig. 4.2e

Fig. 4.2c

Fig. 4.2f

Blow-Out Fractures of Left Orbit

Case Report

Axial and Coronal CT of Orbits and Paranasal Sinuses

Axial CT: There is left proptosis and displacement of the medial orbital wall into the ethmoid sinus and the nasal cavity, with associated soft-tissue extension into the left ethmoids.

The left optic nerve is thickened and the left orbital fat has a higher attenuation than the right.

The left superior orbital fissure is decreased in width, suggesting an associated fracture at this site.

Coronal CT: The left orbital floor is interrupted and fractured, with downward prolapse of soft tissue into the left maxillary antrum. There is further soft-tissue thickening at the lower lateral margin of the left maxillary antrum, which may represent either blood clot or mucosal thickening. The coronal views confirm the enlargement of the left optic nerve.

Diagnosis

The appearances are of post-traumatic left orbital medial and inferior blow-out fractures, with a probable left optic nerve haematoma.

Clinical Data

An 18-year-old man with abdominal pain and diarrhoea.

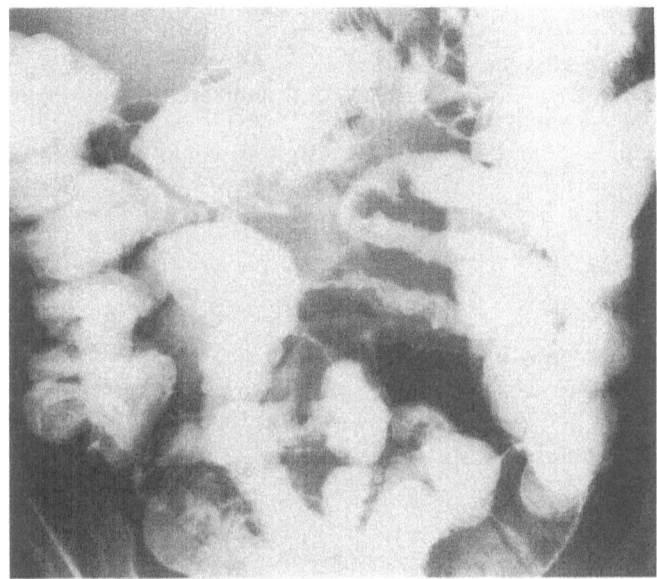

Fig. 4.3a

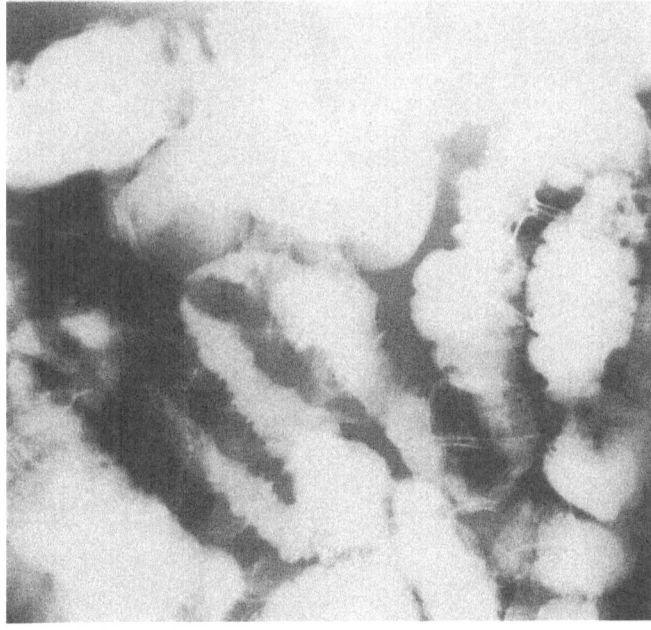

Fig. 4.3b

Intestinal Henoch–Schönlein Purpura

Case Report

Small-bowel Study: There is separation of the small-bowel loops, indicating thickening of the small-bowel wall. The mucosa shows numerous thumbprint-like indentations, predominantly on the mesenteric borders, and the valvulae in these areas are abnormally thick. No ulcers are seen and the small bowel is of normal diameter.

The visualised part of the large bowel is normal.

The appearances are those of a combined mucosal and serosal abnormality affecting the mid part of the small bowel.

The possible differential is long but in a previously well young man is much more limited.

Differential Diagnosis

1. Crohn's disease is always likely in this age group.
2. Intestinal bleeding due to coagulation disorders, trauma, or Henoch–Schönlein purpura.
3. Lymphoma or possibly metastases should be considered with these appearances.

Clinical Data

Female, 25 years old, with vaginal bleeding.

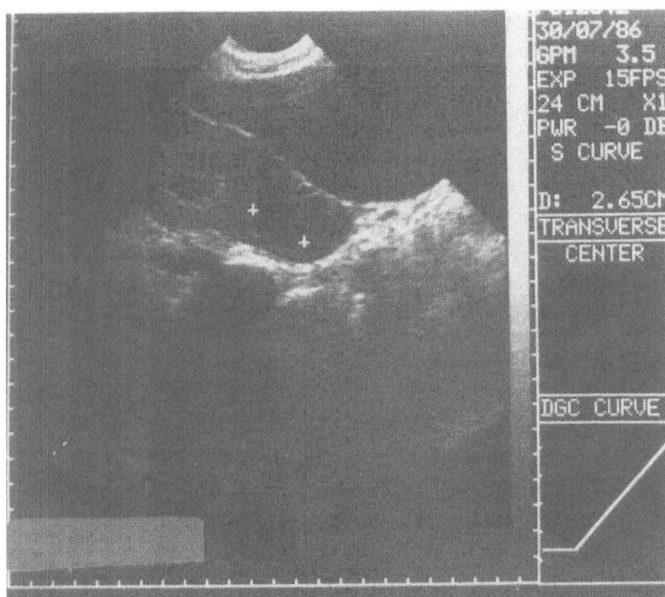

Fig. 4.4a

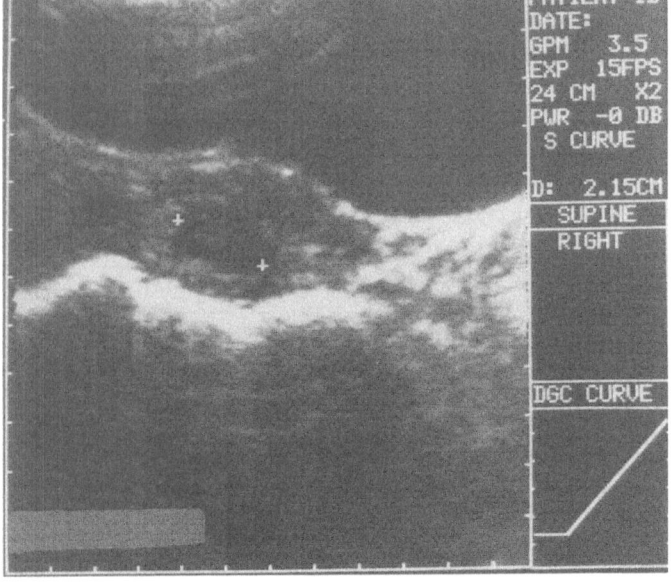

Fig. 4.4b

Ectopic Pregnancy

Case Report

Pelvic Ultrasound: The uterus contains an oval, echo-poor area centrally, surrounded by increased echoes, which represents a gestation sac. There is a further 2.6-cm diameter mass in the position of the left ovary with a low reflectivity, eccentric cystic area within it; this, in view of the above, represents a corpus luteum cyst. There is a third mass (2-cm in diameter) with thick walls and a central cystic area which contains poorly defined echoes.

Differential Diagnosis

This last mass represents either an ectopic gestation sac or a haemorrhagic ovarian cyst. It would be helpful to know if a fetal heartbeat was identified within this mass.

Another possibility for this appearance is a tubo-ovarian abscess.

Clinical Data

Female, 25 years old, with vaginal bleeding.

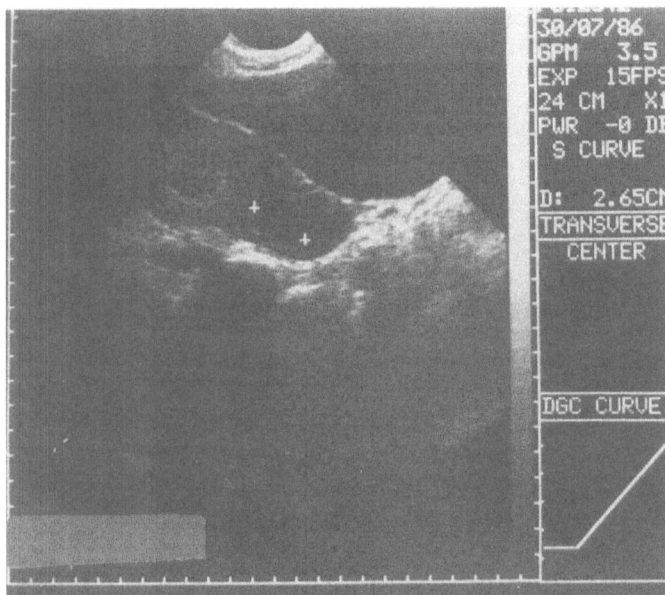

Fig. 4.4a

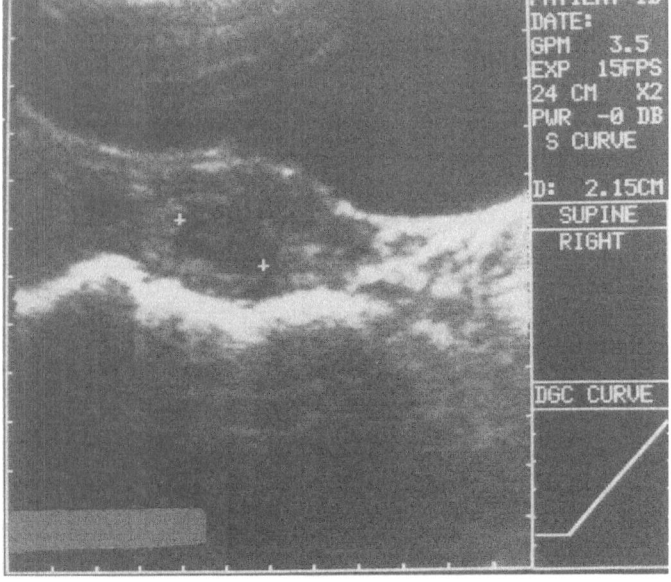

Fig. 4.4b

Ectopic Pregnancy

Case Report

Pelvic Ultrasound: The uterus contains an oval, echo-poor area centrally, surrounded by increased echoes, which represents a gestation sac. There is a further 2.6-cm diameter mass in the position of the left ovary with a low reflectivity, eccentric cystic area within it; this, in view of the above, represents a corpus luteum cyst. There is a third mass (2-cm in diameter) with thick walls and a central cystic area which contains poorly defined echoes.

Differential Diagnosis

This last mass represents either an ectopic gestation sac or a haemorrhagic ovarian cyst. It would be helpful to know if a fetal heartbeat was identified within this mass.

Another possibility for this appearance is a tubo-ovarian abscess.

Clinical Data

A 40-year-old female with swollen hands.

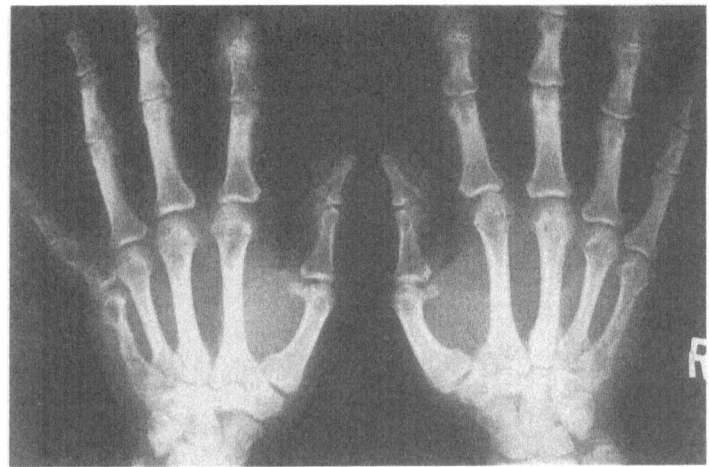

Fig. 4.5a

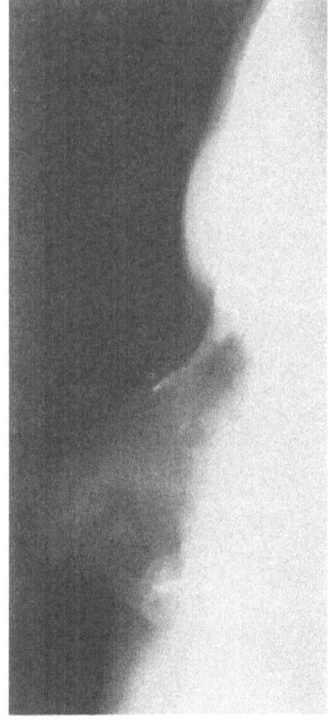

Fig. 4.5b

Sarcoid

Case Report

Both Hands: There are numerous well-defined cystic bony lesions in the phalanges and metacarpals, and the affected bones all have a coarse lacy pattern. There is destruction of three terminal phalanges (acro-osteolysis) with associated soft-tissue swelling.

Nasal Bones: Loss of normal bony architecture is seen at the nasion, with expansion and a coarse trabecular pattern of the nasal bones at this site.

Differential Diagnosis

1. Sarcoid; the hand lesions are characteristic and the nasal bones are affected in this disease.
2. The hand lesions could be due to gout, but this is much less likely.

Clinical Data

Neonate with abdominal distension.

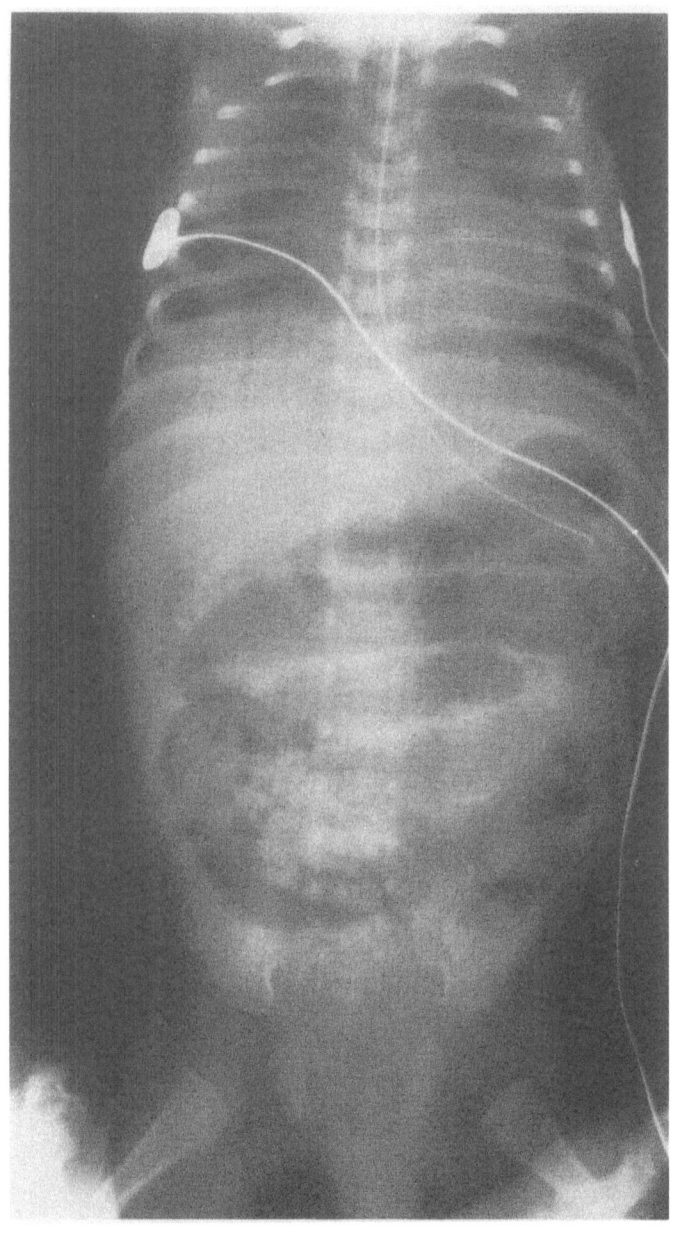

Fig. 4.6

Necrotising Enterocolitis

Case Report

AP Chest: There are some poorly defined areas of patchy density in the right upper zone, right base, and left upper zone. A nasogastric tube passes into the stomach.

AP Abdomen: Several distended loops of thick-walled bowel are visible, and curvilinear lucencies are seen adjacent to the left margin of the spine, extending into the flanks. A circular lucency in the right flank has the appearance of a target lesion. The ill-defined lucencies seen in the periphery of the liver suggest portal-vein gas.

Differential Diagnosis

The appearances are those of intramural gas due to necrotising enterocolitis. Is the patient premature, or is there evidence of Hirschsprung's disease, maternal illness, or intrauterine infection?

Clinical Data

A 50-year-old male with back pain.

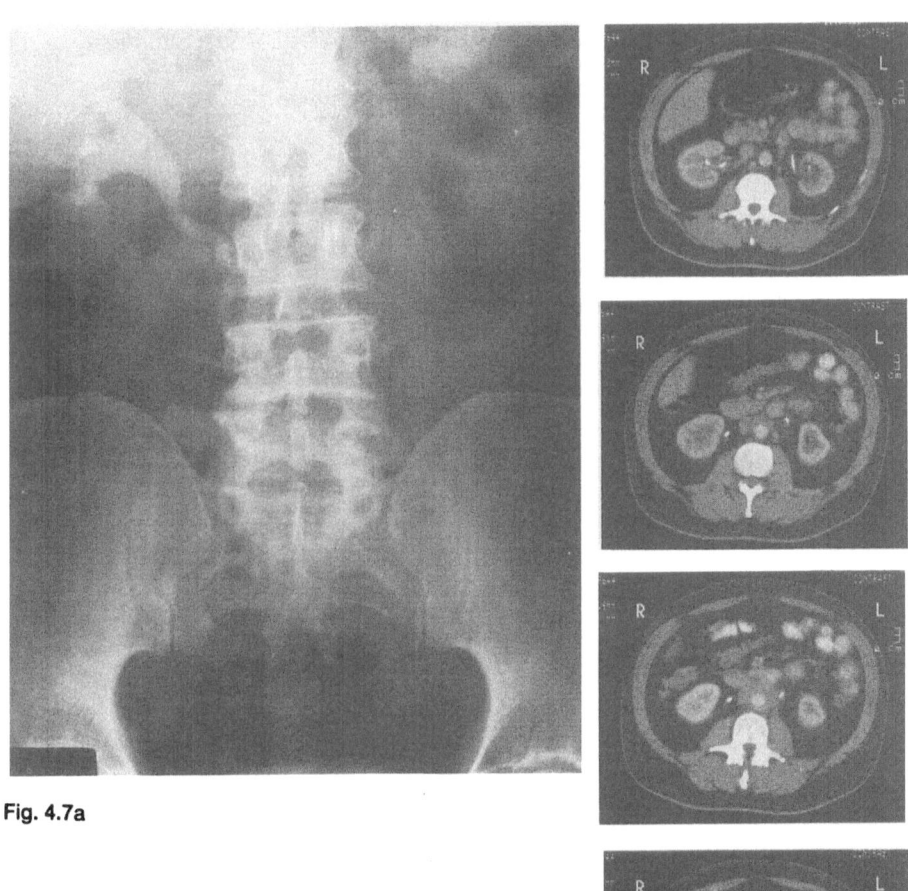

Fig. 4.7a

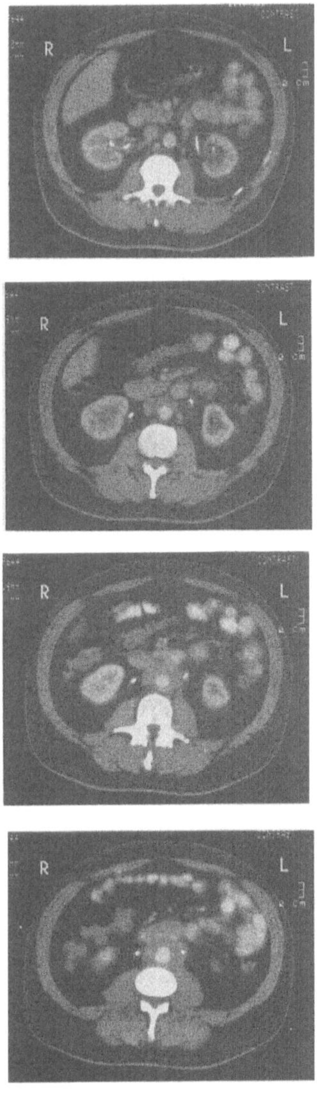

Fig. 4.7b

Idiopathic Retroperitoneal Fibrosis

Case Report

IVU (2-h Full-Length Film): A control film is not available but one should be examined to exclude calcification. There is no significant excretion on the left, and on the right the pelvicaliceal systems are moderately dilated and the upper ureter shows marked medial deviation at the level of L3. Some contrast does reach the bladder.

CT Abdomen: There are high-attenuation densities in both ureters during the early nephrographic phase, indicating that these densities represent stents rather than contrast. The aorta, which is normal in appearance, is surrounded by a cuff of soft tissue, and the left ureter is involved in this periaortic tissue mass, with a minimally dilated left pelvicaliceal system.

The right ureter lies adjacent to the periaortic tissue.

Differential Diagnosis

The marked right upper ureteric displacement is suggestive of retroperitoneal fibrosis, and the appearances around the aorta are more typical of the idiopathic type, although a diffuse lymphoma must be considered.

Retroperitoneal fibrosis secondary to aortic aneurysm is unlikely in the presence of an aorta of normal calibre.

Clinical Data

A 2-year-old girl with wrist pain.

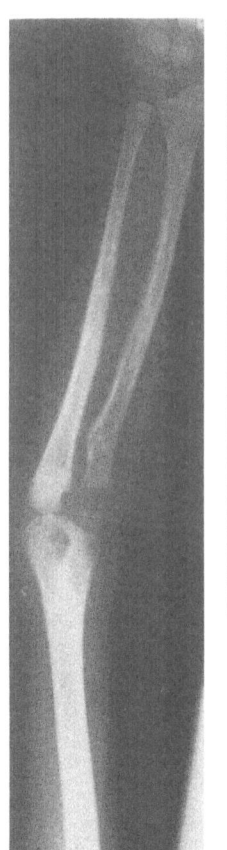

Fig. 4.8a

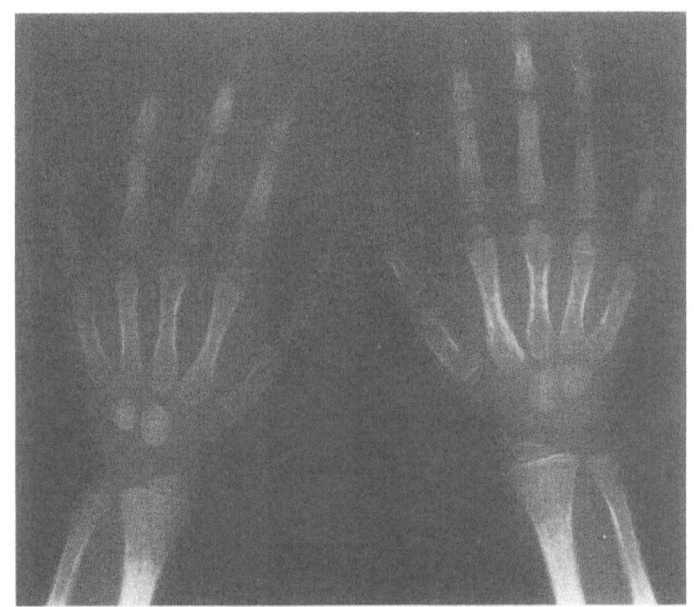

Fig. 4.8b

Leukaemia

Case Report

Right Forearm and Wrist: There is a linear periosteal reaction along the shaft of the humerus and ulna, and the trabecular pattern of the radius is abnormal, with poorly defined lucent areas in the proximal radius. Transverse lucent bands are present at the distal ulna and radial metaphyses.

Both Hands: There are punched-out destructive lesions in both 5th metacarpals, with further small cystic lesions within the metacarpal shafts. Periosteal reactions are seen along the shafts of many of the phalanges and metacarpals.

Differential Diagnosis

1. Leukaemia results in all the above findings and is the most likely cause, although neuroblastoma may also produce these features.
2. Congenital syphilis can also produce these appearances, but is very rare.

Examination 5

Clinical Data

Female, 18 years old, with a pyrexia.

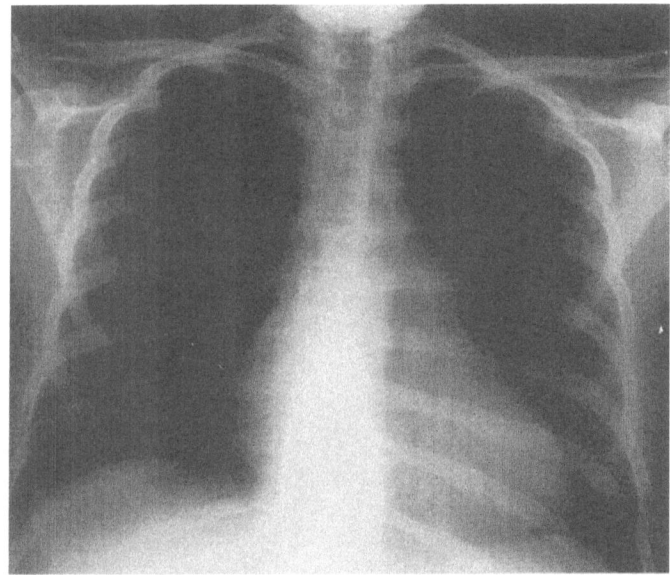

Fig. 5.1a

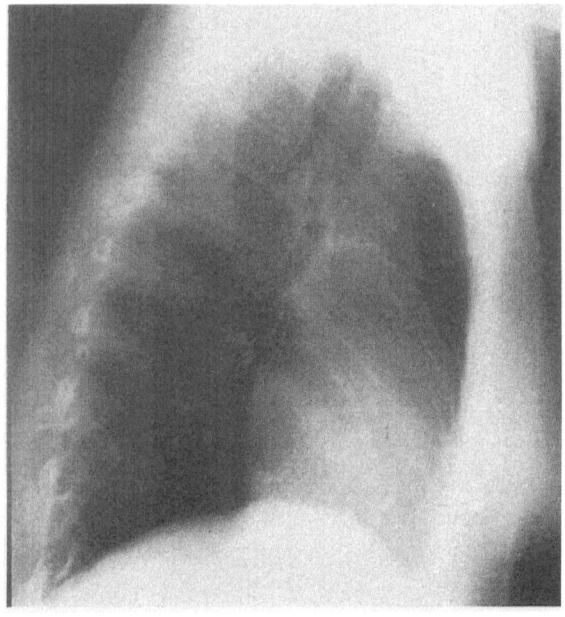

Fig. 5.1b

Right Atrial Myxoma

Case Report

PA Chest: The heart is mildly enlarged and the lungs are normal. There is coarse, dense calcification projected through the heart, which, on the *Lateral Film*, is seen to lie within the heart in the region of the right atrium.

Differential Diagnosis

1. Myxoma of the right atrium.
2. Calcified thrombus ball lying in right atrium.
3. A calcified infarct is less likely, since the calcification appears to lie within the cardiac chamber.

Clinical Data

A 3-year-old boy with abdominal pain.

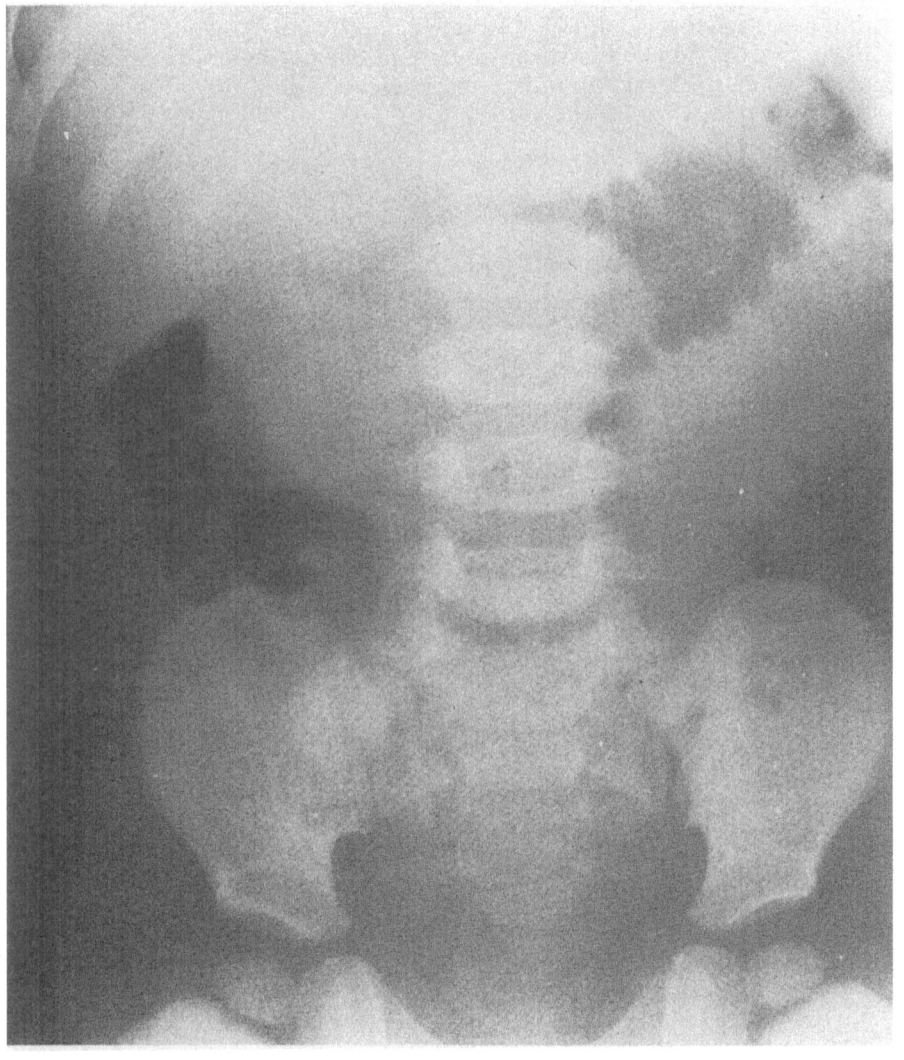

Fig. 5.2a

Intussusception

Case Report

Plain Abdomen: A rounded, soft-tissue mass projects into the lumen of the transverse colon in mid-abdomen. The rectum is empty. There is no evidence of dilated small-bowel loops.

Diagnosis

The appearances are those of an intussusception affecting the transverse colon, with no associated small-bowel obstruction.

If the clinical situation is stable, a barium enema should be carried out to confirm the diagnosis and to reduce the intussusception.

Supplementary Film

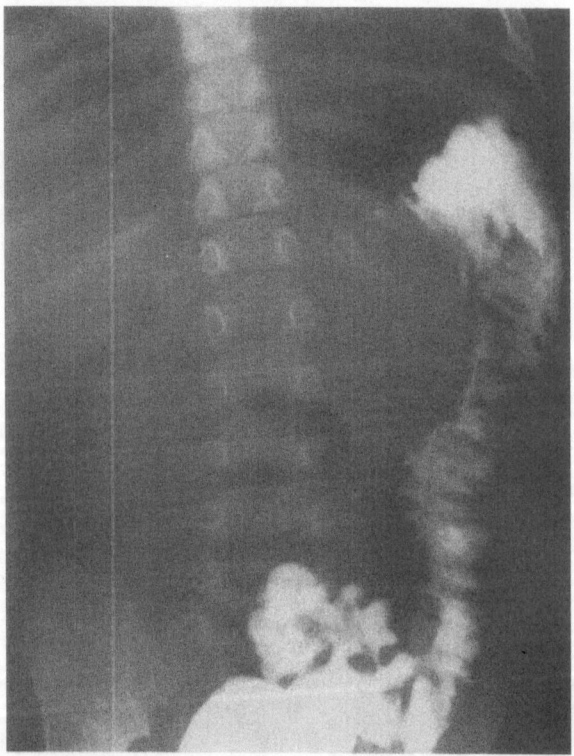

Fig. 5.2b

Barium Enema: *This shows a large filling defect with a coiled-spring appearance in the transverse colon typical of intussusception.*

Clinical Data

A 3-year-old girl who is asymptomatic and is being followed up by the paediatric clinic.

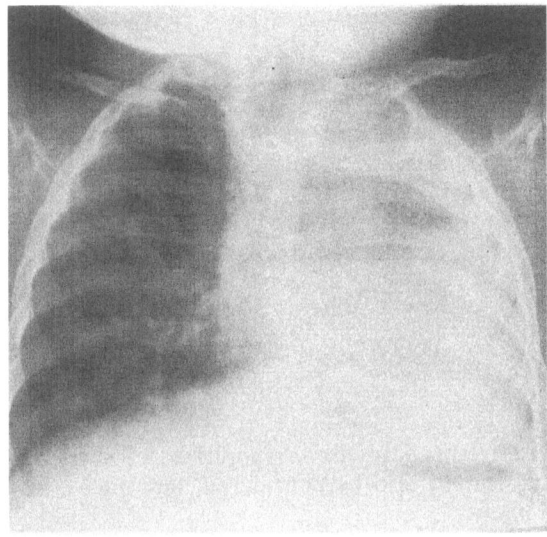

Fig. 5.3a

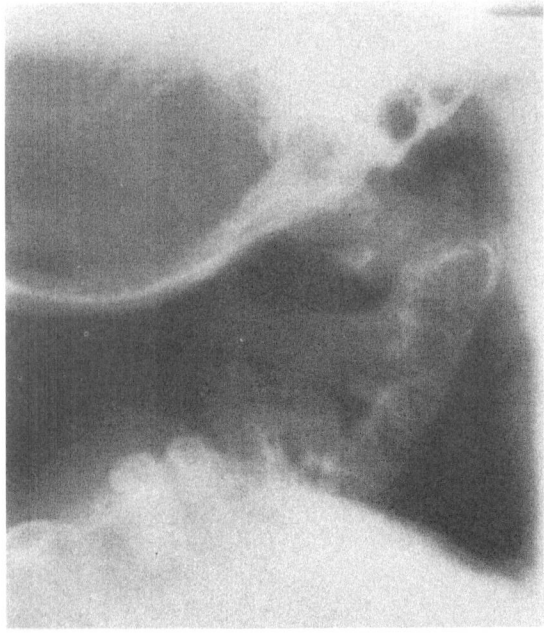

Fig. 5.3b

Hypoplastic Left Lung, Treated Fallots, Klippel–Feil, Cervical Spina Bifida, and Aplastic Odontoid Peg

Case 5.3

Case Report

PA Chest: The left hemithorax is completely opaque and there is shift of the mediastinum to the left. The left rib interspaces are closer together than the right, confirming marked volume loss. Sternal sutures are visible, suggesting previous cardiac surgery.

The cervical spine is abnormal, with widening of the interpedicular distance and failure of fusion of the lower cervical laminae.

Lateral Cervical Spine: All the visible vertebral bodies are fused. The laminae are grossly disorganised, with a combination of vertical fusions and aplasia. The odontoid peg is absent and the anterior arch of the atlas is displaced posteriorly relative to C2. The skull base is not adequately visualised to determine if platybasia is present.

Differential Diagnosis

The appearances on the PA Chest are probably of a hypoplastic left lung, but complete collapse of the left lung should also be considered, although in the absence of symptoms this is less likely.

The appearances of the cervical spine suggest:

1. Cervical spina bifida.
2. Klippel–Feil type abnormality of upper cervical spine.
3. Aplasia of the odontoid peg and posterior displacement of C1 relative to C2. Careful flexion and extension views should be obtained to assess stability.

Clinical Data

A 6-year-old girl who fell over in the playground.

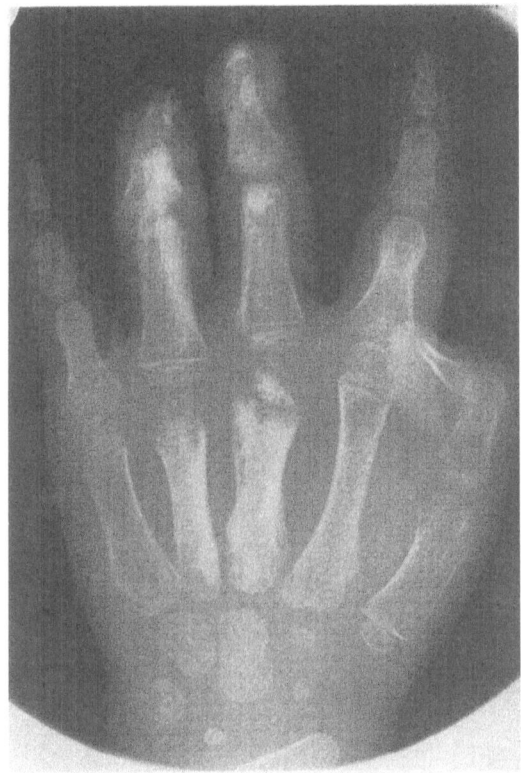

Fig. 5.4a

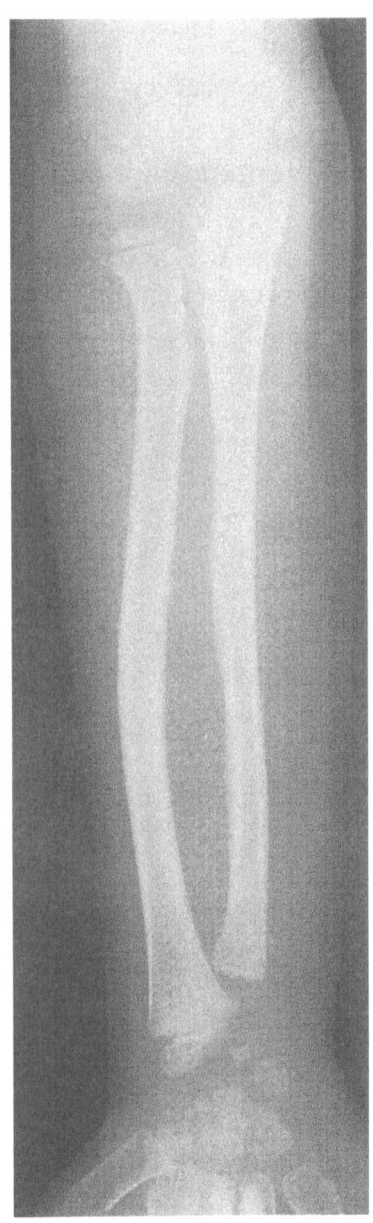

Fig. 5.4b

Melorheostosis

Case Report

Hand and Lower Arm: There is sclerosis of the radial side of the fourth metacarpal and phalanges, and of the ulnar side of the third metacarpal and phalanges. Sclerosis of the proximal ulna is also present.

The bone modelling is normal but the ulna is abnormally short. No lucencies are visible. Enquiry should be made about the possibility of other bone involvement.

The abnormalities are in a ray or sclerotomal distribution.

Differential Diagnosis

1. Melorheostosis is the most likely cause for this pattern of changes.
2. Fibrous dysplasia should also be considered, but the distribution is unusual.

Clinical Data

Female, 43 years old, with chronic dyspnoea.

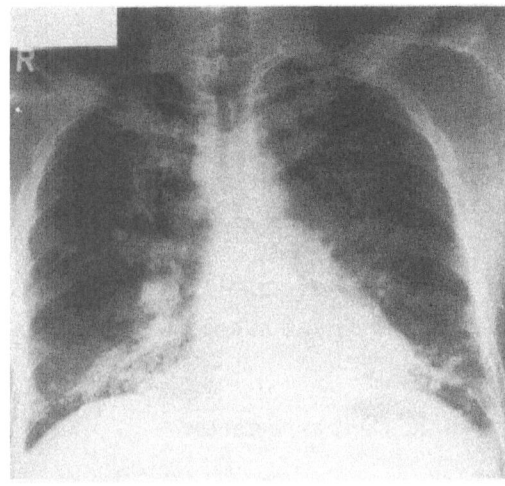

Fig. 5.5a

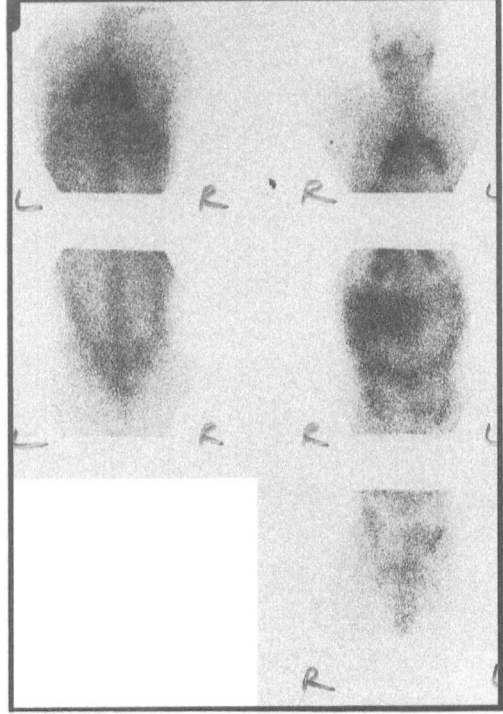

Fig. 5.5b

Sarcoid

Case Report

PA Chest: Both hila are grossly enlarged and lobulated, the right being larger than the left. The heart size is normal.

Bilateral, relatively symmetrical linear, and poorly defined nodular opacities are present in all lung zones. These are confluent in the mid and lower zones.

Gallium Scan: There is a marked increase of uptake in both lungs, with a perihilar distribution; there is also a slight increase in activity in the nose, both parotid glands and the lacrimal glands.

Differential Diagnosis

1. Sarcoid is the most likely cause of all the above findings.
2. Lymphoma would also be possible, but the chest film appearances are very severe for this condition.
3. Tuberculosis is possible.
4. Berylliosis is rare, but could cause all the chest appearances.

Clinical Data

A 40-year-old male with abdominal pain and constipation.

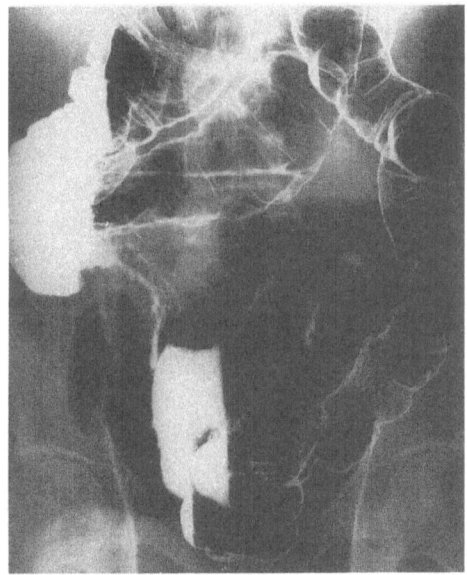

Fig. 5.6a

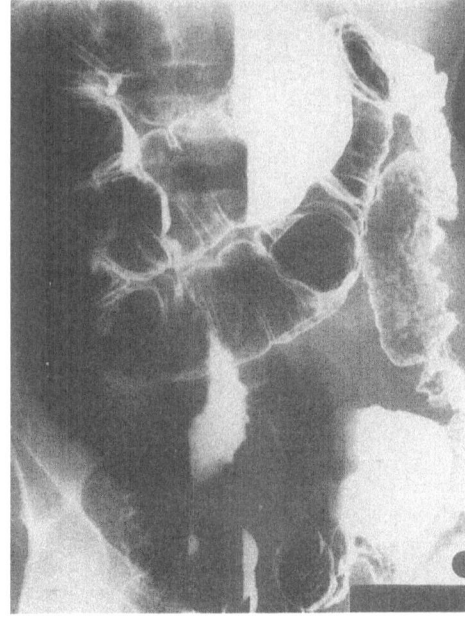

Fig. 5.6b

Crohn's Colitis

Case Report

Double-Contrast Barium Enema: There are at least three different sites of abnormality in the descending and sigmoid colon, separated by normal-looking bowel. Aphthous ulcers are seen in the transverse and descending colon, and there is a 6-cm long stricture in the distal ileum which shows deep ulceration and a cobblestone type mucosal pattern. There are further ulcers in the sigmoid colon. Neither the terminal ileum nor the sacroiliac joints can be seen clearly.

Differential Diagnosis

1. Crohn's colitis is the most likely cause for these findings, and a small-bowel study should be undertaken.
2. Ischaemic colitis.
3. Behçets disease, infection with Yersinia, and amoebiasis can all produce aphthous ulceration, but are less likely to cause stricture formation.

Clinical Data

A 3-year-old boy with retention.

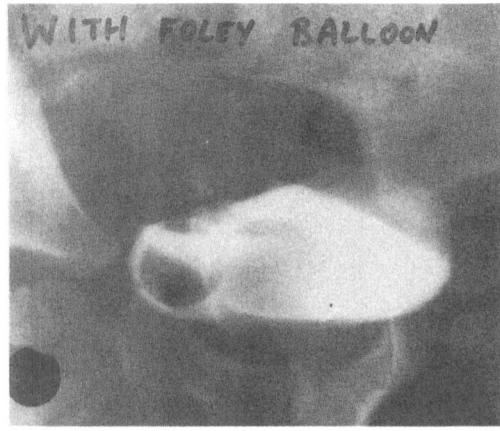

Fig. 5.7a

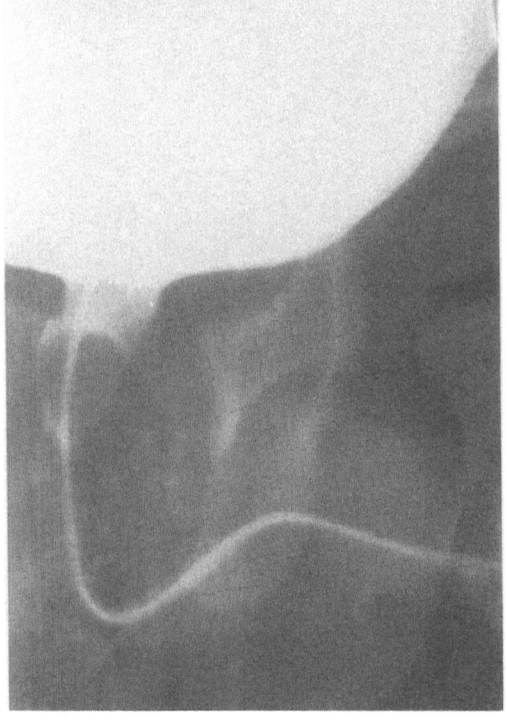

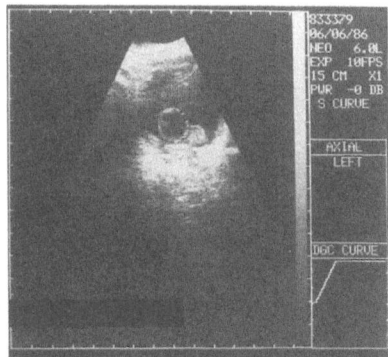

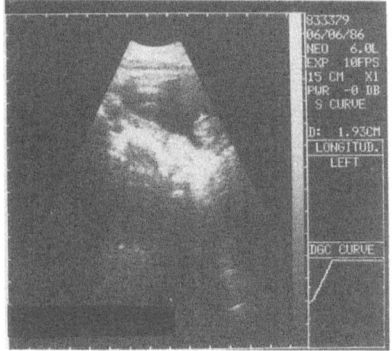

Fig. 5.7b

Fig. 5.7c

Hamartomatous Bladder Polyp

Case Report

Micturating Cystogram: There are two large filling defects within the bladder, one of which is the balloon of a Foley catheter. The pathological filling defect is well defined and is quite mobile, since it prolapses into the posterior urethra during micturition, causing obstruction and preventing the remainder of the urethra from filling.

Bladder Ultrasound: There is a rounded, 1.9-cm echogenic soft-tissue mass arising from the bladder, which does not cast an acoustic shadow; a Foley catheter balloon is visible adjacent to this on the axial view.

Differential Diagnosis

1. Benign polyp.
2. Rhabdomyosarcoma is the most important condition to exclude, but is not usually so well defined.
3. Blood clot is possible.
4. Ureterocele and calculus are largely excluded by the ultrasound.

Clinical Data

A 60-year-old female with pain in the right eye.

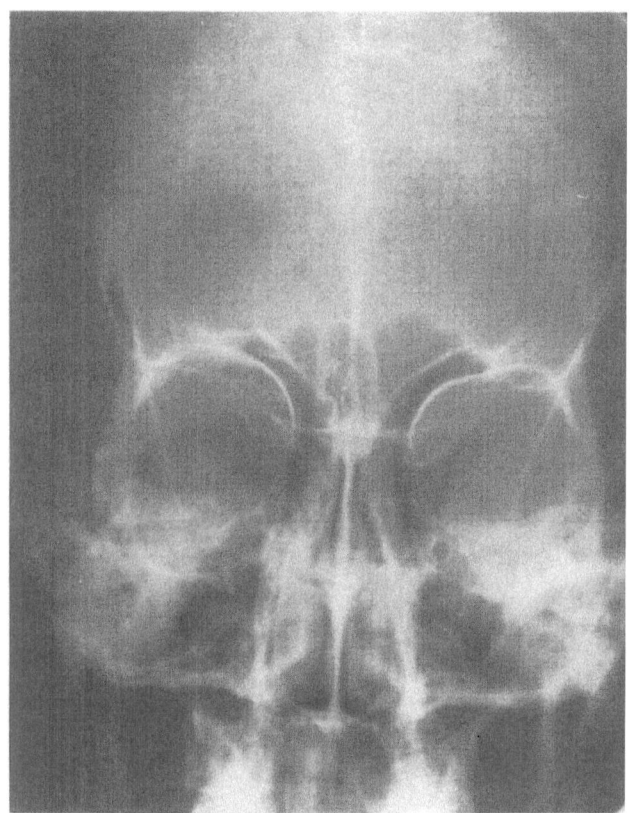

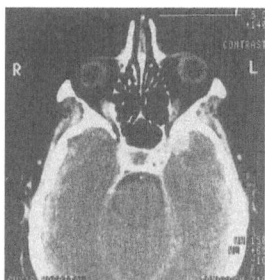

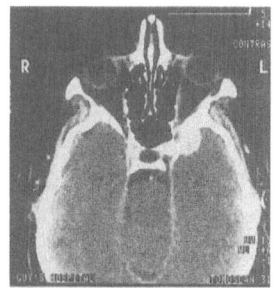

Fig. 5.8a

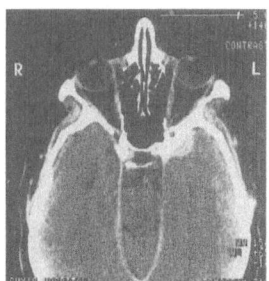

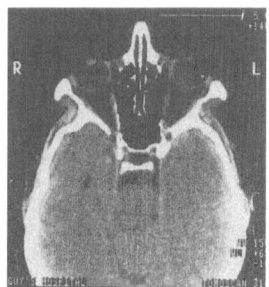

Fig. 5.8b

Left Sphenoidal Ridge Meningioma

Case Report

Skull AP: There is increased density of the greater wing of the left sphenoid seen through the left orbit. The left superior orbital fissure is much more clearly seen than the right, reflecting the increased density of the surrounding bone.

CT Head: The greater wing of the left sphenoid is thickened and there is a dense ovoid area of enhancing tissue lying adjacent to this bone and extending medially towards the cavernous sinus and the optic canal.

Diagnosis

The appearances are those of a meningioma causing hyperostosis of the greater wing of the sphenoid.

Examination 6

Examination 5

Clinical Data

A 50-year-old male with chronic dyspnoea.

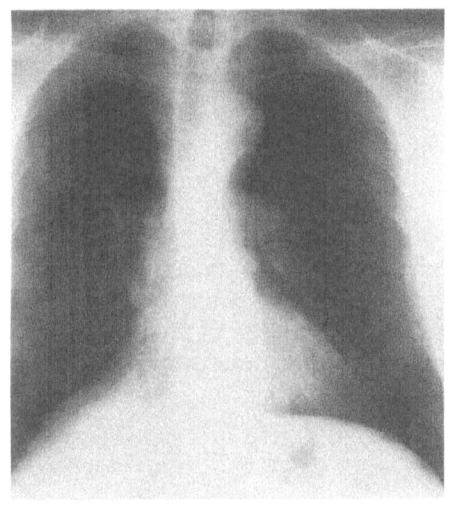

Fig. 6.1a

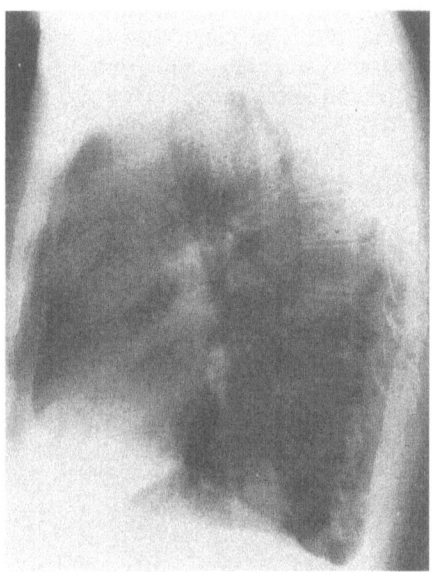

Fig. 6.1b

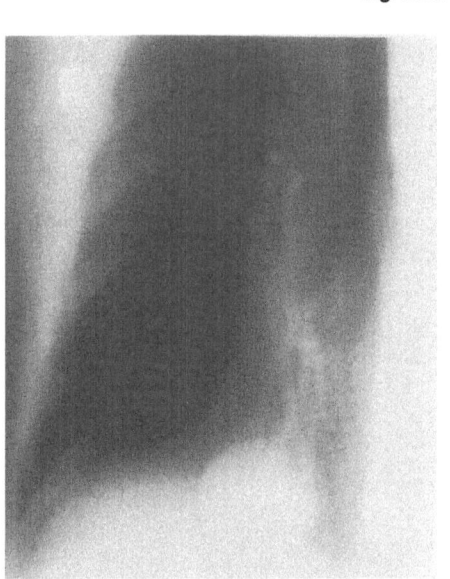

Fig. 6.1c

Rounded Atelectasis (Blesovsky Tumour)

Case Report

PA Chest: The heart size is normal. A round mass is projected under the right hemidiaphragm, which lies in the right lower lobe. There is loss of volume in the right lower zone, indicated by the very medial position of the descending pulmonary artery. No other lung abnormality is present.

Lateral Chest: This confirms the posteroinferior position of the spherical mass which lies on the right hemidiaphragm.

Tomogram This clearly shows a circular mass with an irregular upper margin. Several vessels are seen running into the upper edge of this mass, forming a comet-tail appearance.

Differential Diagnosis

1. The appearances are suggestive of rounded atelectasis. Is there any history of asbestos exposure?
2. Other lung masses such as bronchial carcinoma, abscesses, AVM, etc., should be considered, but are less likely given the above characteristic appearances.

Clinical Data

An 18-year-old female with diarrhoea.

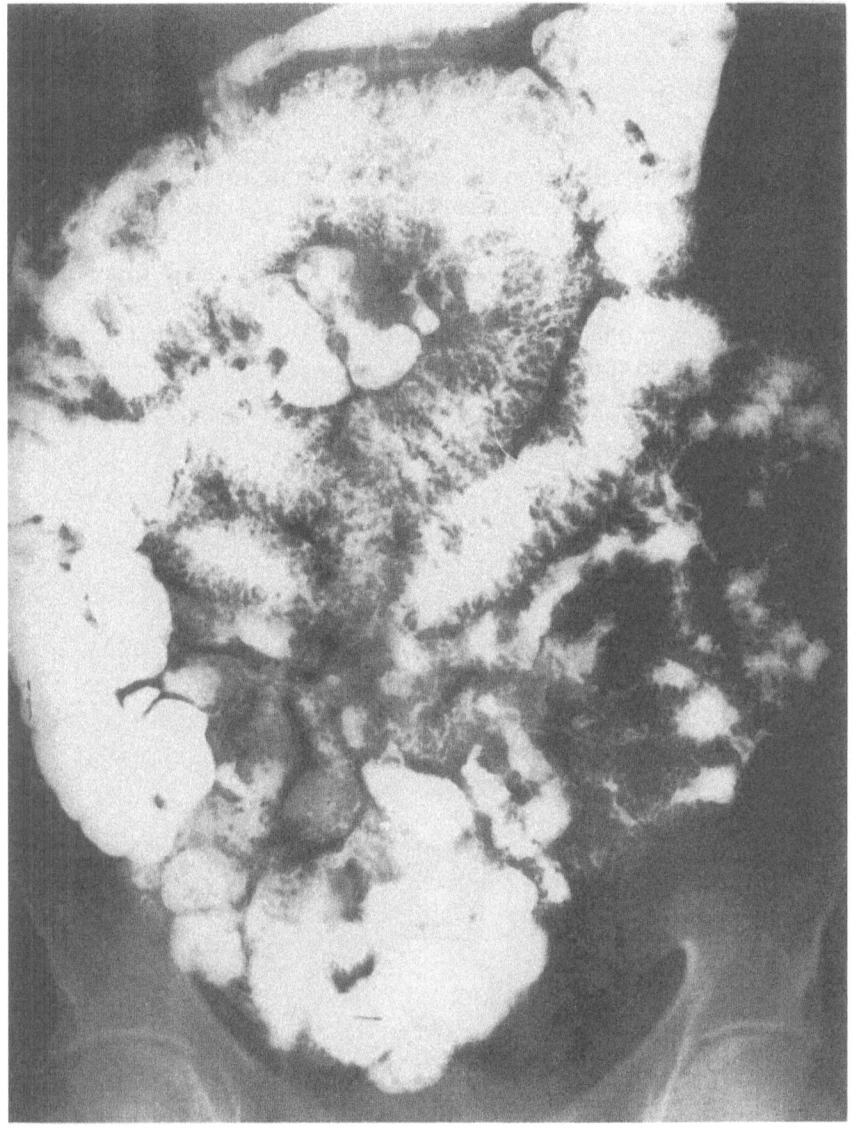

Fig. 6.2

95

Nodular Lymphoid Hyperplasia

Case Report

Small-Bowel Meal: Multiple small nodules are distributed throughout the visible bowel. The nodules are all less than 0.5 cm in diameter and too numerous to count. Some of the loops are underfilled and further views are required to confirm that these areas distend normally.

The folds are of normal thickness and the diameter of the small bowel is normal. No strictures are seen.

Differential Diagnosis

1. Nodular lymphoid hyperplasia is the most likely cause for these appearances, and clinical history of gut infections and IgA and IgM deficiency should be sought.
2. Lymphoma can look similar but not as diffuse as this, and is therefore less likely.
3. Whipple's disease and mastocytosis are not so widespread and are much less likely causes for this appearance.

Clinical Data

A 70-year-old female with bloody diarrhoea.

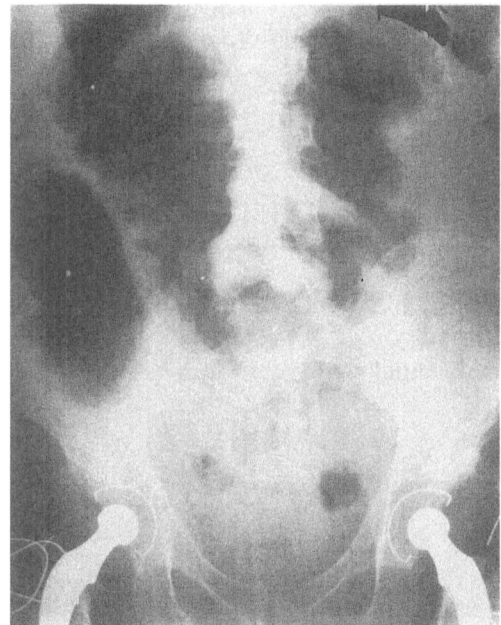

Fig. 6.3a

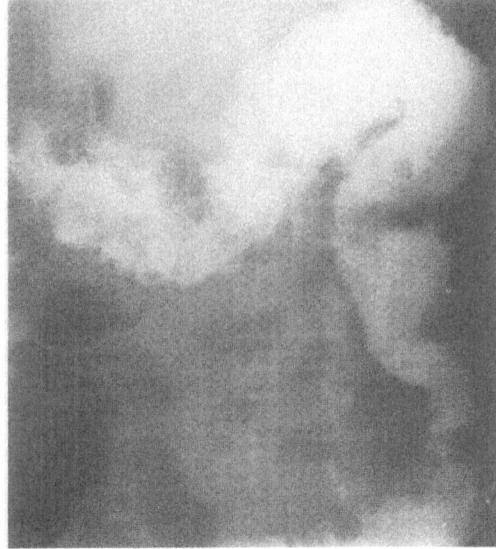

Fig. 6.3b

Pseudomembranous Enterocolitis

Case Report

Plain Abdomen: Bilateral total hip replacements are present. The ascending and transverse colon are dilated, and the visible bowel wall shows multiple, thick, thumbprint-like soft-tissue indentations along its length.

Single-Contrast Enema: This confirms the numerous indentations into the bowel lumen, which represent areas of mucosal oedema. No ulcers or strictures are seen; the abnormality affects both sides of the bowel wall and is continuous.

Differential Diagnosis

1. Antibiotic-associated pseudomembranous enterocolitis is quite likely, bearing in mind the bilateral hip replacements.
2. Ulcerative colitis with toxic dilatation is also likely, but the lack of ulcers on the enema is a little against this possibility.
3. Ischaemic colitis is also possible, but is unlikely to be so widespread.

Clinical Data

An 8-year-old boy with ataxia.

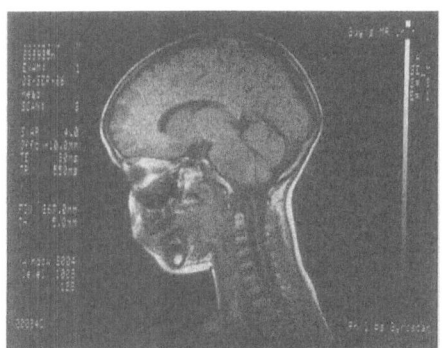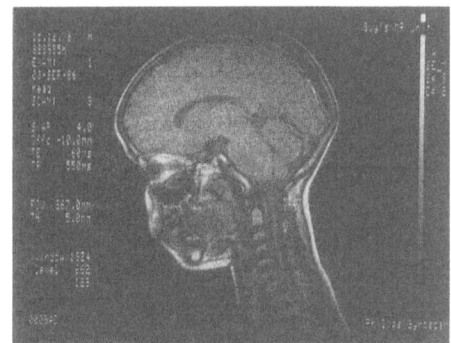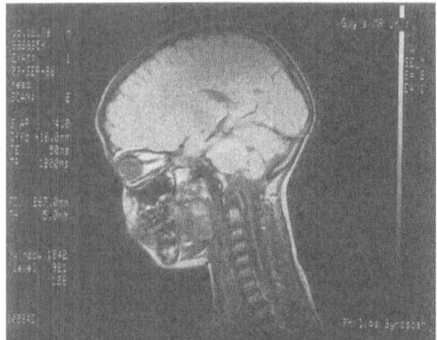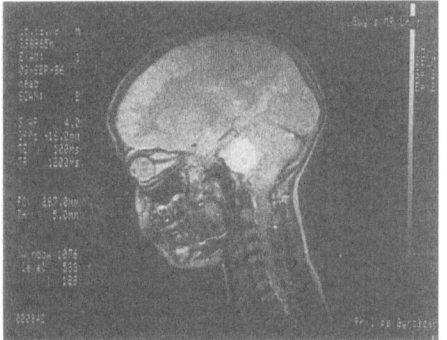

Fig. 6.4

Brain-Stem Glioma

Case Report

Sagittal Head Magnetic Resonance Images: T1 and T2 weighted sequences are shown. The brain stem is expanded and contains an area of high signal intensity. The ventricles are not enlarged and the cerebellum is indented and displaced posteriorly.

Diagnosis

The appearances are of a mass lesion within the brain stem, which is probably a glioma.

Clinical Data

A 50-year-old male involved in a road traffic accident.

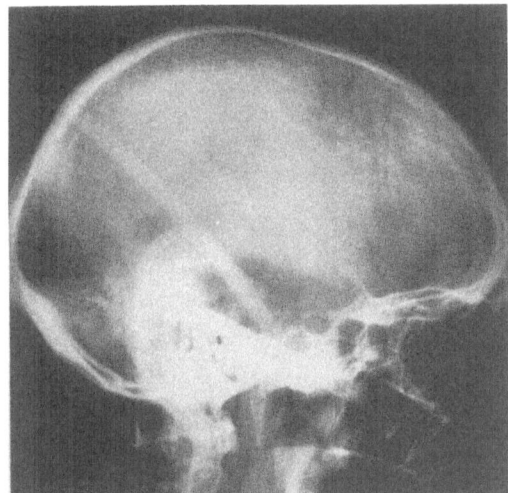

Fig. 6.5a

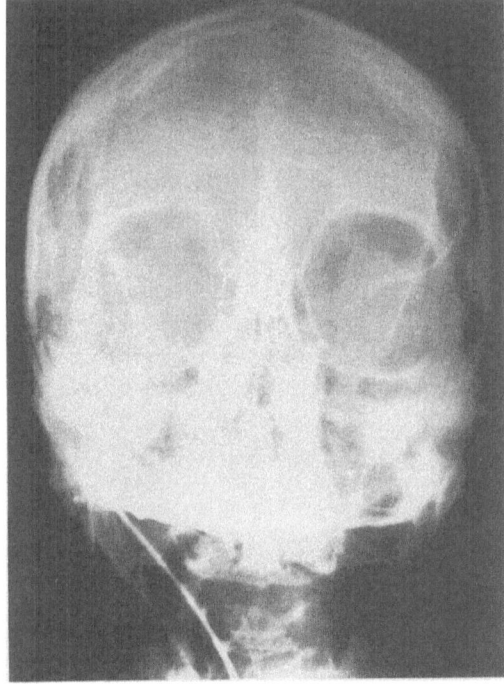

Fig. 6.5b

Multiple Skull and Cervical Spine Trauma

Case Report

Lateral Skull: There is an air–fluid level in the sphenoid sinus, indicating a fracture at the base of the skull. An oblique branching lucency in the posterior parietal region represents a complex vault fracture.

AP Skull: There is an elevated fractured segment in the right parietal region. There is also an oblique fracture of the left articular mass of C2, and the relationship of the right lateral mass of the atlas to the peg and to the articular surface of C2 is abnormal. The right C1–C2 joint space is decreased, with medial displacement of the right lateral mass of C1, suggesting a fracture of the lamina of C1.

Diagnosis

Multiple skull and upper cervical spine fractures.

Clinical Data

A 3-year-old boy with recurrent urinary-tract infections.

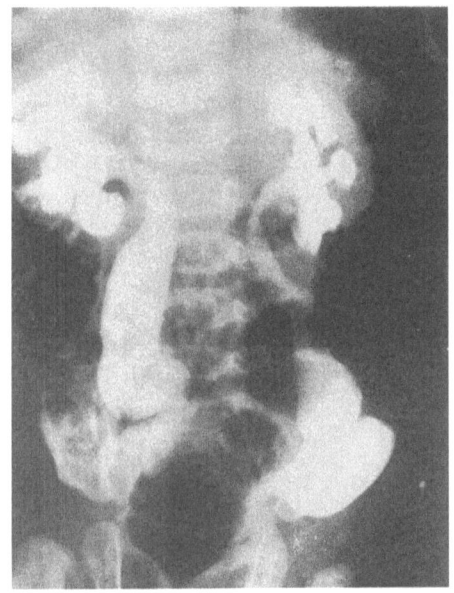

Fig. 6.6a

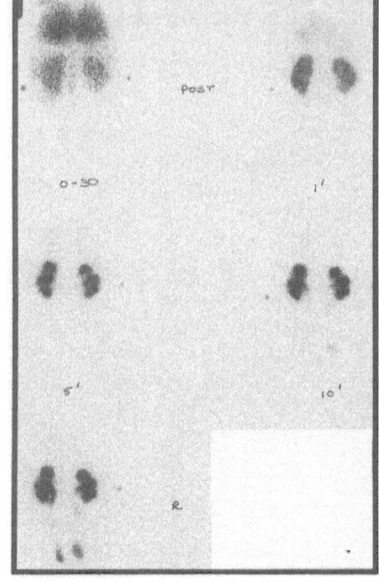

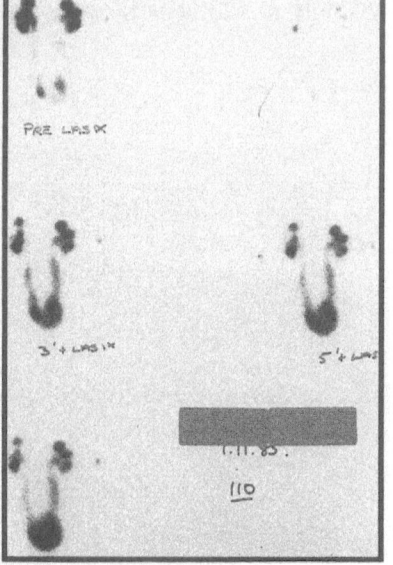

Fig. 6.6b   Fig. 6.6c

103

Neuropathic Bladder

Case Report

IVU (Full Length Film): No control film available. Both pelvicaliceal systems are markedly dilated and the calices are clubbed. The renal outlines are not well seen, but there is some parenchymal loss of the left upper pole. The ureters are also both markedly dilated and the bladder is large.

The acetabula are shallow and further films are required to assess the possibility of hip dislocation.

DTPA Scan: There is symmetrical perfusion and function of both kidneys, with a slow transit time through dilated pelvicaliceal systems and ureters. After frusemide, there is drainage from both renal pelves into dilated ureters, and although no quantitation is available there is probably more than 50% washout.

Differential Diagnosis

1. Neuropathic bladder with upper-tract dilatation.
2. Severe bilateral reflux.
3. Urethral valves.
4. Prune-belly syndrome.
5. Pelviureteric junction obstruction is ruled out by the DTPA scan, but bilateral vesicoureteric junction obstruction cannot be excluded.

Micturating cystogram and urodynamic study, plain films of lumbosacral spine, and possibly a Whitaker (flow perfusion) test are required for further assessment.

Clinical Data

A 30-year-old male with pain in the left elbow.

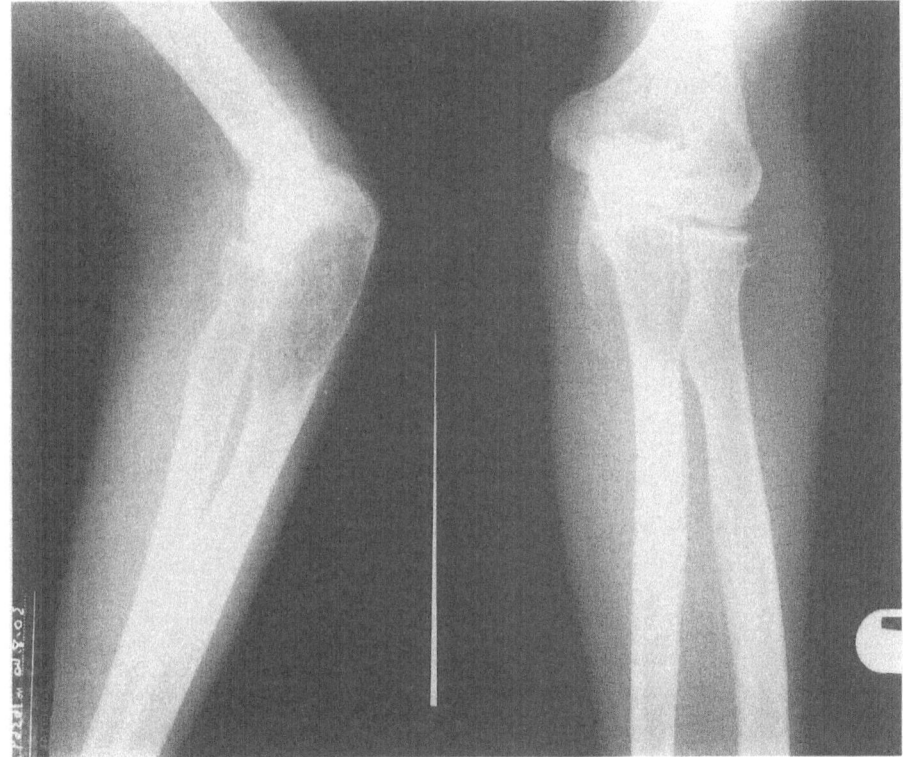

Fig. 6.7a

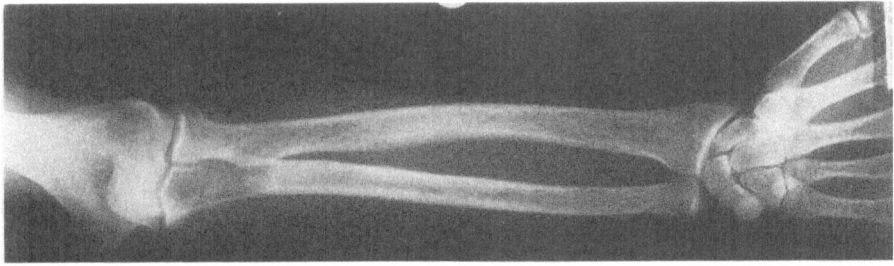

Fig. 6.7b

Brown Tumour

Case Report

Left Elbow: There is a large lucent area in the left olecranon extending up to the joint margin, causing slight expansion of the ulna. The lesion has a wide zone of transition, contains thick septa, and lies centrally within the bony metaphysis.

AP View: This shows minimal periosteal reaction but no soft-tissue mass is seen. The radius and ulna have a coarse trabecular pattern, possibly due to disuse osteoporosis.

Differential Diagnosis

1. Giant-cell tumour.
2. Brown tumour.
3. Aneurysmal bone cyst.
4. Metastases (much less likely in this age group).

Clinical Data

A 4-year-old male.

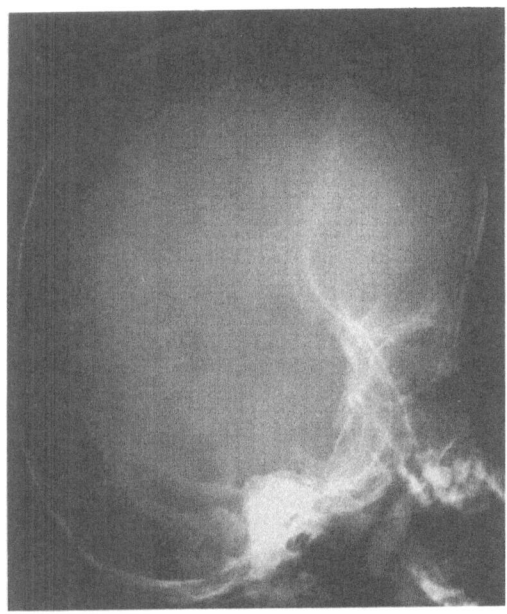

Fig. 6.8a

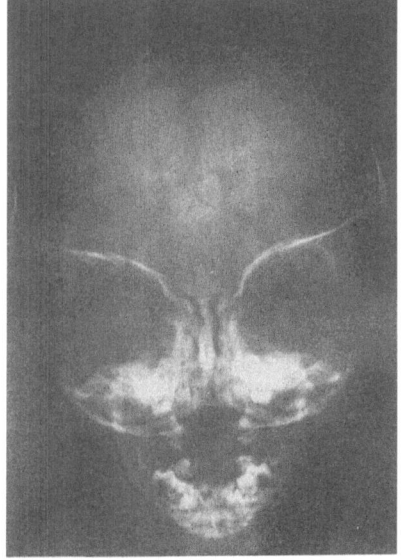

Fig. 6.8b

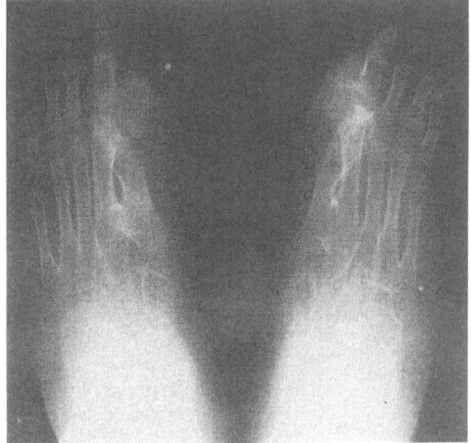

Fig. 6.8c

Apert's Syndrome (Acrocephalosyndactyly)

Case Report

Lateral and AP Skull: The skull shape is very abnormal, with a marked increase in the height of the vault and a relative shortening in the A–P plane (turricephaly). The lambdoid and sagittal sutures are clearly seen, but the coronal sutures are absent, suggesting that they are fused.

There is hypertelorism, and the maxilla is hypoplastic.

Both Feet: There is partial bony fusion between the 1st and 2nd metacarpals bilaterally, and deformity of the phalanges of both great toes, which are abnormally broad.

Differential Diagnosis

1. Apert's syndrome (acrocephalosyndactyly); the above changes are most typical of this condition.
2. Carpenter's syndrome (acrocephalopolysyndactyly) is possible, but this is usually associated with polydactyly.

Examination 7

Clinical Data

A 4-year-old boy, coughing and unwell.

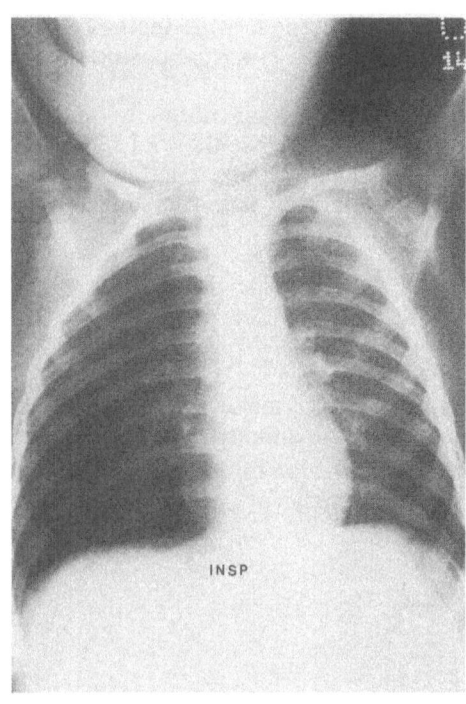

Fig. 7.1a

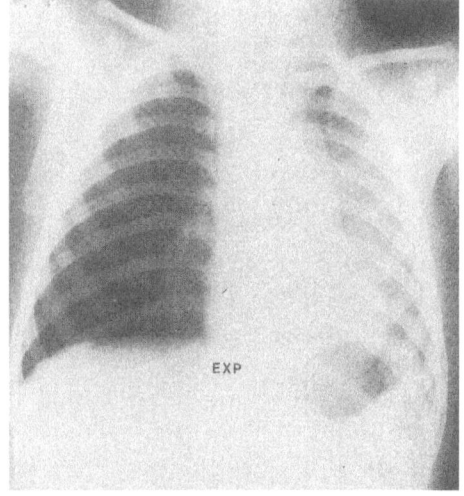

Fig. 7.1b

Inhaled Foreign Body

Case Report

PA Chest, Inspiratory and Expiratory

Inspiratory Film: Despite some rotation there is a significant difference in the lucency of the lungs. The vascular pattern of the left hemithorax is normal, whereas the right lung vascularity appears diminished, with increased lung volume and increased lucency. The mediastinum is in a normal position.

Expiratory Film: The right lung has failed to deflate normally and the mediastinum is now swung to the left. The appearances are those of right-sided obstructive emphysema with an increased right lung volume. The failure of the right lung to deflate in expiration confirms that this is the abnormal lung and that the asymmetry is not due to a dense left lung.

Differential Diagnosis

1. An obstructive lesion in the right main bronchus producing a ball-valve effect. A foreign body or bronchial adenoma are the most likely.
2. McLeod's syndrome or pulmonary artery agenesis are unlikely, since the right lung volume is large and the patient is symptomatic.

   Bronchoscopy is indicated.

Clinical Data

A 6-year-old girl with fits.

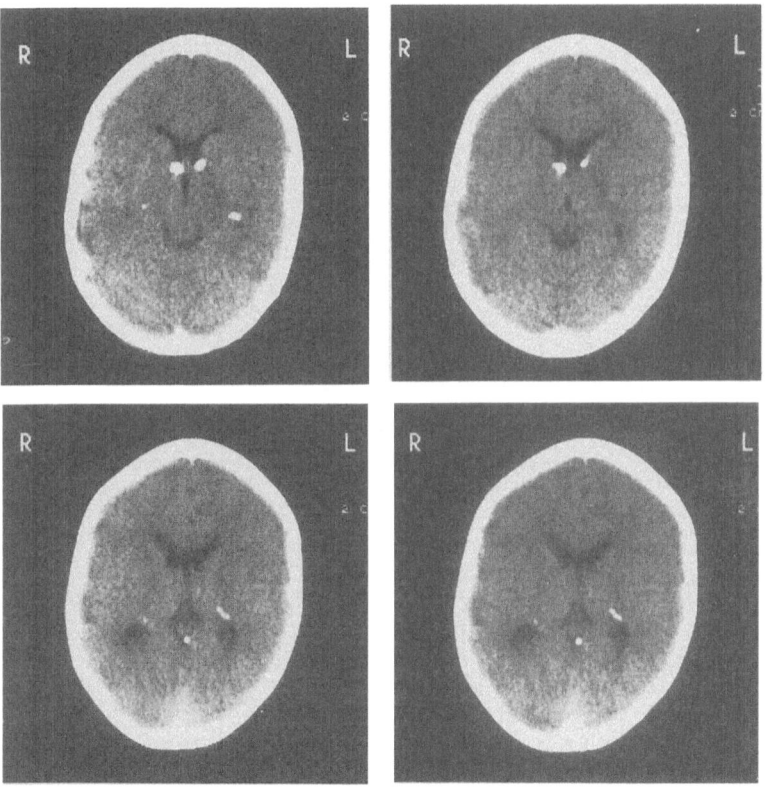

Fig. 7.2

Tuberose Sclerosis

Case Report

Head CT: Bilateral, circular, small calcific nodules are visible in a periventricular distribution, particularly around the trigones of the lateral ventricles. No mass effect is visible and the ventricles are normal in size. No other abnormality is seen.

Differential Diagnosis

1. Tuberose sclerosis is the most likely cause for these appearances.
2. Intrauterine infection by toxoplasmosis or cytomegalovirus may cause calcification in a similar distribution and should also be considered.

Clinical Data

Female, 34 years old, investigated for an eosinophilia.

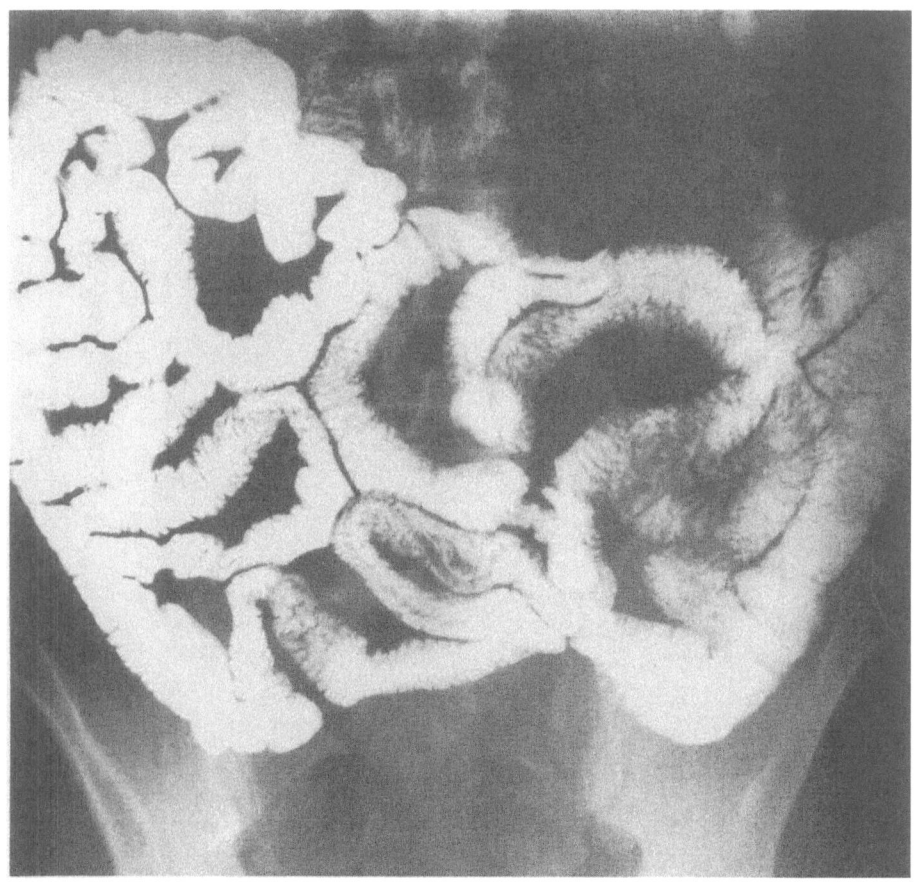

Fig. 7.3

Small-Bowel Ascariasis

Case Report

Small-Bowel Meal: There is a long, lucent filling defect in the small bowel.

Diagnosis

The appearances are those of a parasite such as ascaris.

Clinical Data

Woman, 56 years old, with clubbing.

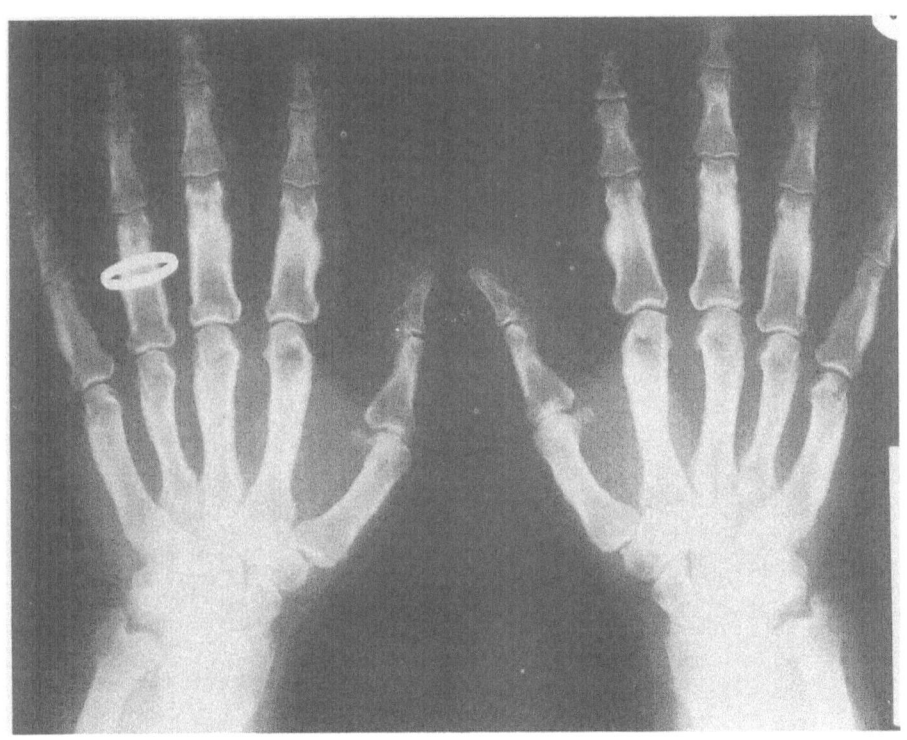

Fig. 7.4

117

Thyroid Acropachy

Case Report

Both Hands: There is a thick, well-organised periosteal reaction involving the radial aspects of the proximal phalanges of the index fingers, the ulnar aspects of the remainder of the phalanges, and both 5th metacarpals. The distal radius and ulna are normal.

Differential Diagnosis

1. Thyroid acropachy is the most likely cause for this appearance. Is there a history of thyroid disease?
2. Hypertrophic osteoarthropathy usually produces a more laminar periosteal reaction, but could cause these findings.
3. Pachydermoperiostitis commonly causes a thickened cortex, but occurs almost exclusively in males.

Clinical Data

Male, 55 years old, with abdominal pain.

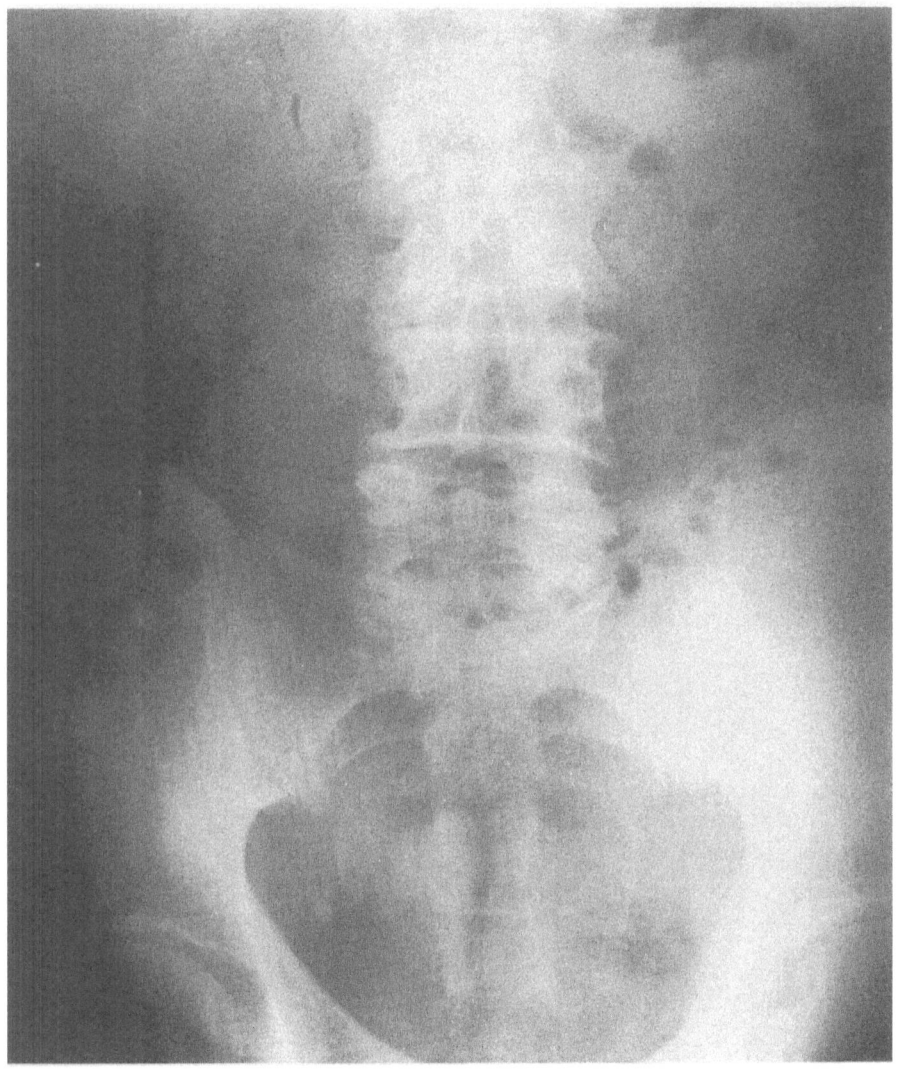

Fig. 7.5

119

Bowel Obstruction With Fluid-Filled Loops

Case Report

Plain Abdomen: The abdomen is very dense overall and few gas-filled loops are visible, but numerous small, circular gas lucencies are present in the mid-abdomen. No gas is visible in the rectum.

The film demonstrates the "string of beads" sign which is indicative of bowel obstruction, but with largely fluid-filled rather than solely gas-filled loops. Abdominal ultrasound will confirm the fluid-filled loops and show whether peristalsis is present.

Diagnosis

Bowel obstruction or paralytic ileus.

Clinical Data

Female, 56 years old.

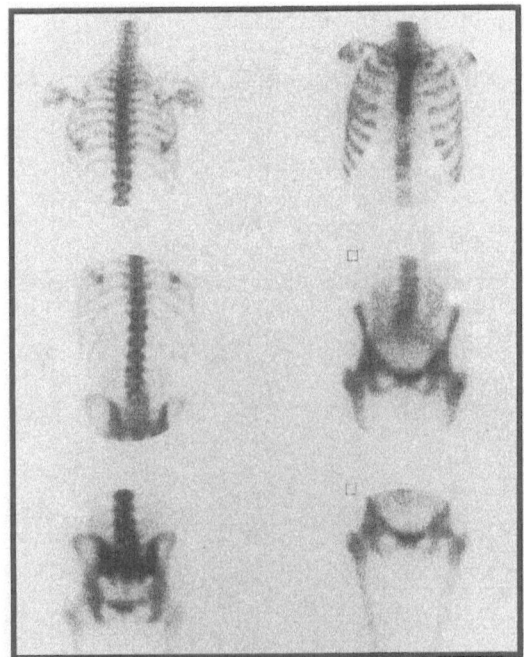

Fig. 7.6a

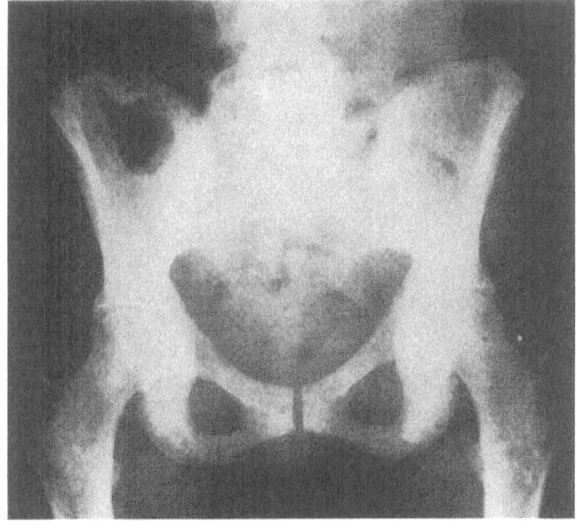

Fig. 7.6b

Breast Carcinoma Metastases

Case Report

Bone Scan: There is very good definition of all bones seen. There is slightly increased uptake in the right femoral head, but no other focal areas of abnormal uptake. The soft-tissue uptake is minimal and only faint renal excretion is visible in low-lying kidneys. The appearances are those of a "superscan", which may be due either to metabolic bone disease or to diffuse metastases.

AP Pelvis There is diffuse patchy sclerosis throughout all the bones and no normal trabeculae are visible.

Diagnosis

The appearances are those of widespread metastases most probably from breast carcinoma.

Clinical Data

Pregnancy of 34 weeks, referred for possible anomaly.

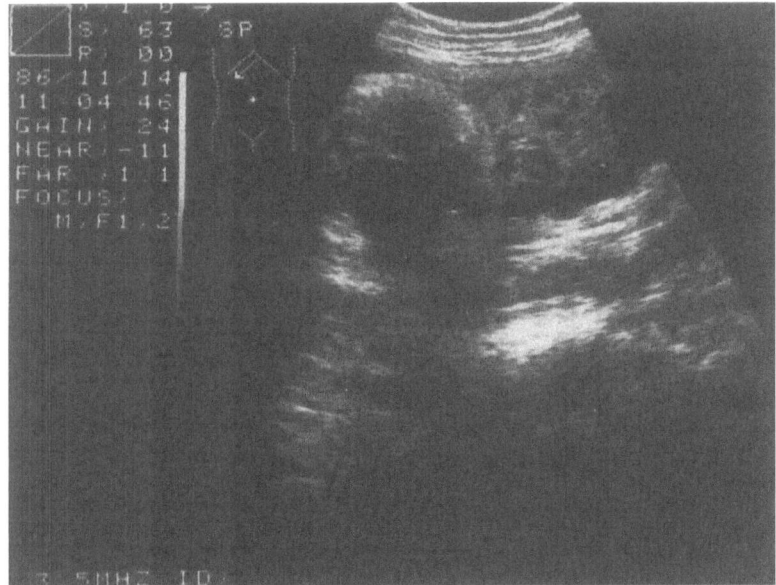

Fig. 7.7a

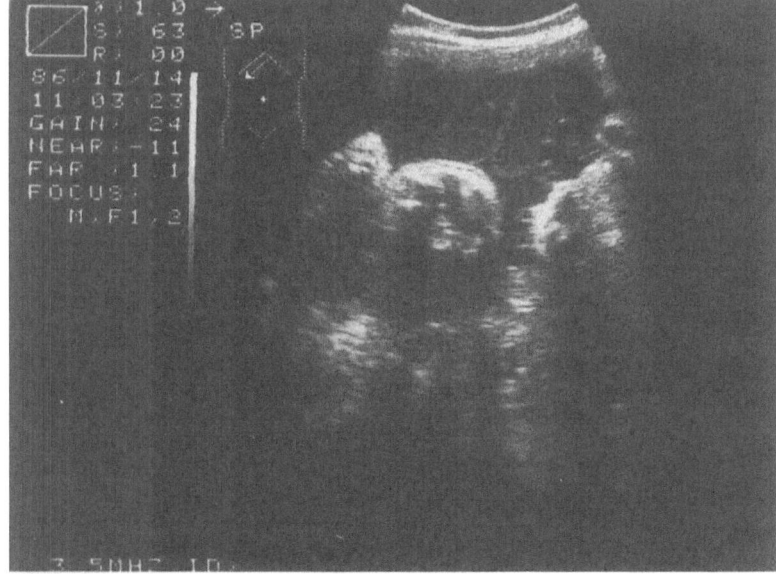

Fig. 7.7b

Gastroschisis/Omphalocele

Case Report

Fetal Ultrasound (Longitudinal and Transverse Scans): A complex cystic and echogenic mass is seen arising from the abdomen in the transverse section. The longitudinal view shows well-defined oval echogenic rings with sonolucent centres. These have the appearances of bowel loops in amniotic fluid.

The umbilical vessels are not clearly seen and no membrane is identified around the complex mass.

Differential Diagnosis

The appearances are those of an anterior abdominal-wall defect, either omphalocele or gastroschisis. The lack of a covering membrane is more in keeping with the latter diagnosis.

Detailed ultrasound of the remainder of the fetus is indicated, since both abnormalities are associated with other anomalies.

Clinical Data

Female, 55 years old, with right proptosis.

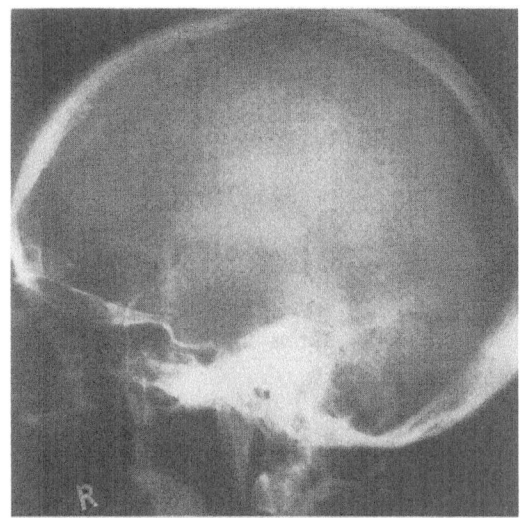

Fig. 7.8a

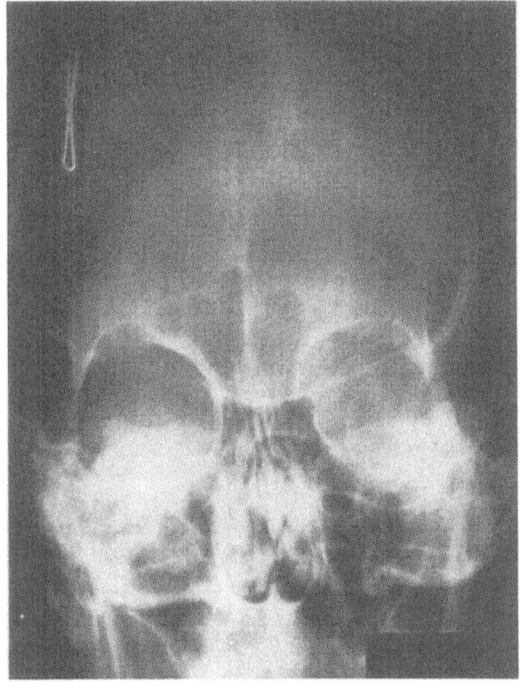

Fig. 7.8b

Neurofibromatosis

Case Report

Lateral Skull: A large, well-defined lytic defect which does not have a sclerotic rim is projected over the frontal region adjacent to the squamous temporal fissure. No other vault lesion is visible.

AP Skull: The right sphenoid ridge cannot be seen and there is some expansion of the superior orbital margin. The left orbit is normal. The appearances are of a right-sided bare orbit. A head CT would define the abnormal areas more clearly.

Differential Diagnosis

1. Neurofibromatosis typically produces the combination of a well-defined vault defect and a bare orbit.
2. Metastases could also produce these abnormalities and must be strongly considered in this age group.
3. A purely lytic meningioma may cause a bare orbit but is unlikely to cause the vault defect as well.

Supplementary Film

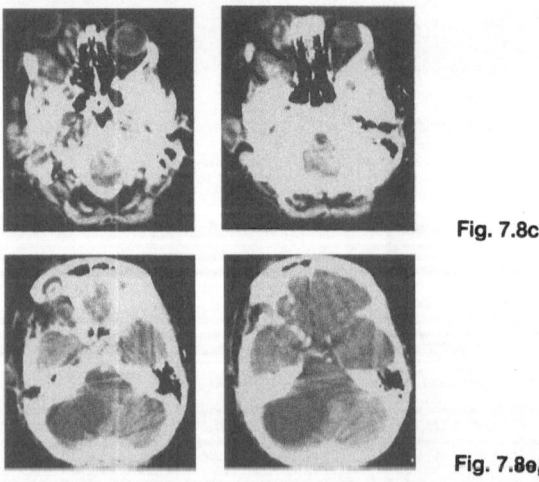

Fig. 7.8c,d

Fig. 7.8e,f

Head CT: *There is right-sided proptosis plus hypoplasia of the right lateral orbital wall and the right greater wing of the sphenoid. The right temporal lobe is prolapsing into the right orbit, causing the proptosis. There is also a large, low-attenuation posterior fossa mass on the right, causing some mass effect. The appearances are of central neurofibromatosis complicated by the development of a posterior fossa glioma.*

126

Examination 8

Clinical Data

A 12-year-old girl with anaemia.

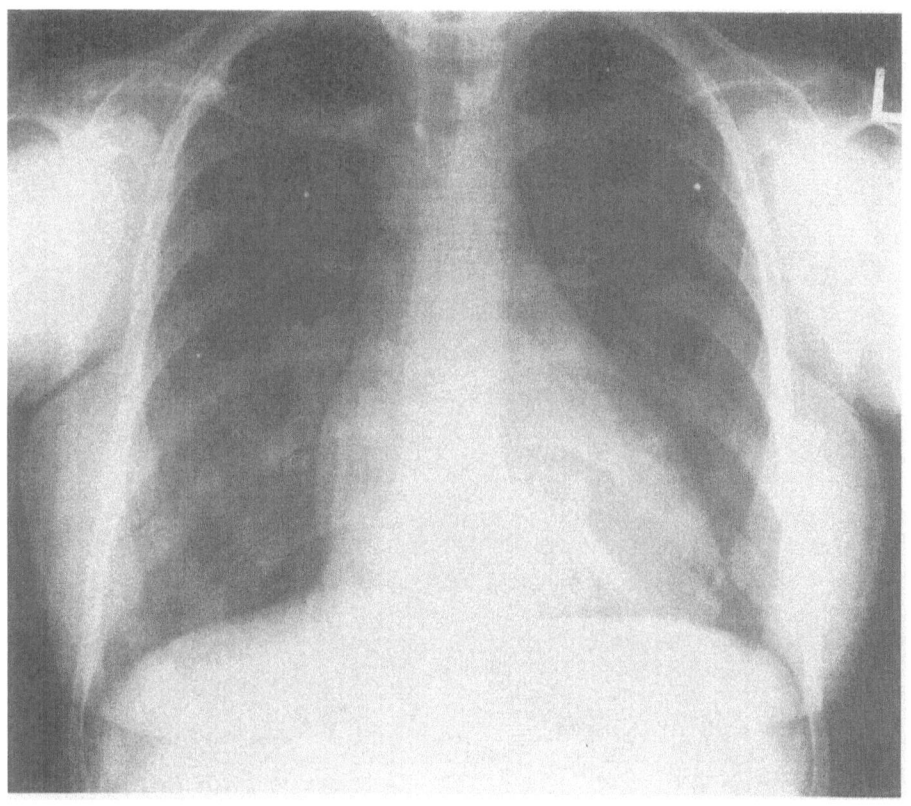

Fig. 8.1

Renal Osteodystrophy

Case Report

PA Chest: The heart is enlarged and the lungs are normal in appearance. Prominent, irregular, large erosions are present, affecting the lateral ends of both clavicles at the acromioclavicular joints. In addition, both humeral heads have a very angulated orientation relative to their shafts, suggesting that there is some epiphyseal slip. Further specific views of the shoulders would be useful to assess this.

Differential Diagnosis

1. Hyperparathyroidism is a probable cause for these appearances, and is likely to be secondary to renal failure in this age group, forming part of the pattern of renal osteodystrophy.
2. Juvenile chronic arthritis can cause these appearances and is also quite likely.
3. Other rare causes of this appearance are cleidocranial dysostosis, scleroderma, progeria, and pyknodysostosis, but all these are very unlikely in comparison to the above.

Clinical Data

Female, 55 years old, with incontinence.

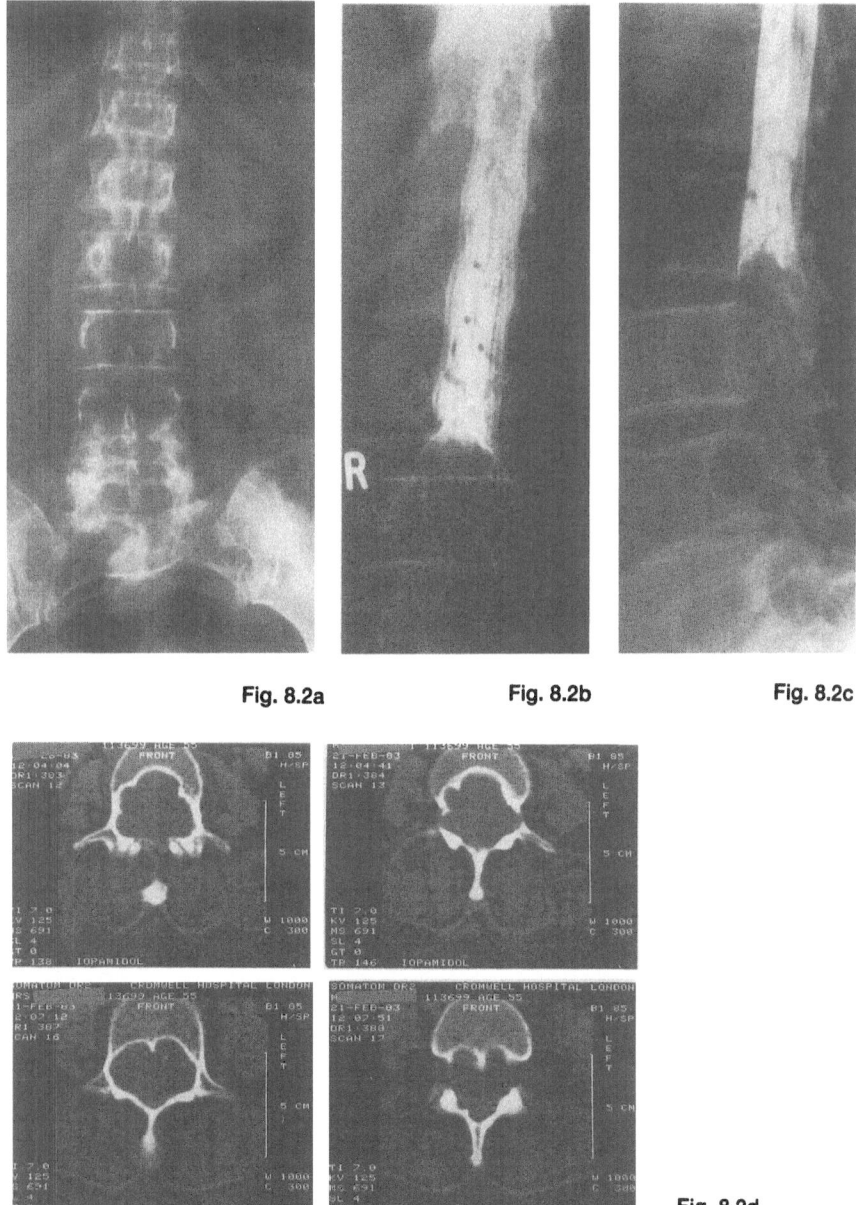

Fig. 8.2a   Fig. 8.2b   Fig. 8.2c

Fig. 8.2d

Lumbosacral Ependymoma

Case Report

AP Lumbar Spine: There is marked widening of the interpedicular distances at the level of L3 and L4, with thinning of the pedicles medially. The laminae are not seen and there is loss of density of the bodies of L3 and L4.

AP and Lateral Myelogram: Marked posterior scalloping of the bodies of L3 and L4 is seen and there is complete block to contrast flow at the level of the L2 disc. An intradural lesion is shown, apparently arising from the nerve roots, widening the thecal sac, and displacing the contrast-filled subarachnoid space. Dilated vessels are seen indenting the thecal sac.

CT Lumbar Spine: A large soft-tissue-density lesion lies within the spinal canal, eroding the laminae, pedicles, and the posterior aspects of the vertebral bodies.

Differential Diagnosis

1. Spinal ependymoma is most probable.
2. Other spinal gliomas are possible but much less likely.
3. Neurofibroma or meningioma.
4. Lymphoma secondaries or metastases are less likely as these are usually extradural.

Clinical Data

A 2-month-old boy with dysmorphic features.

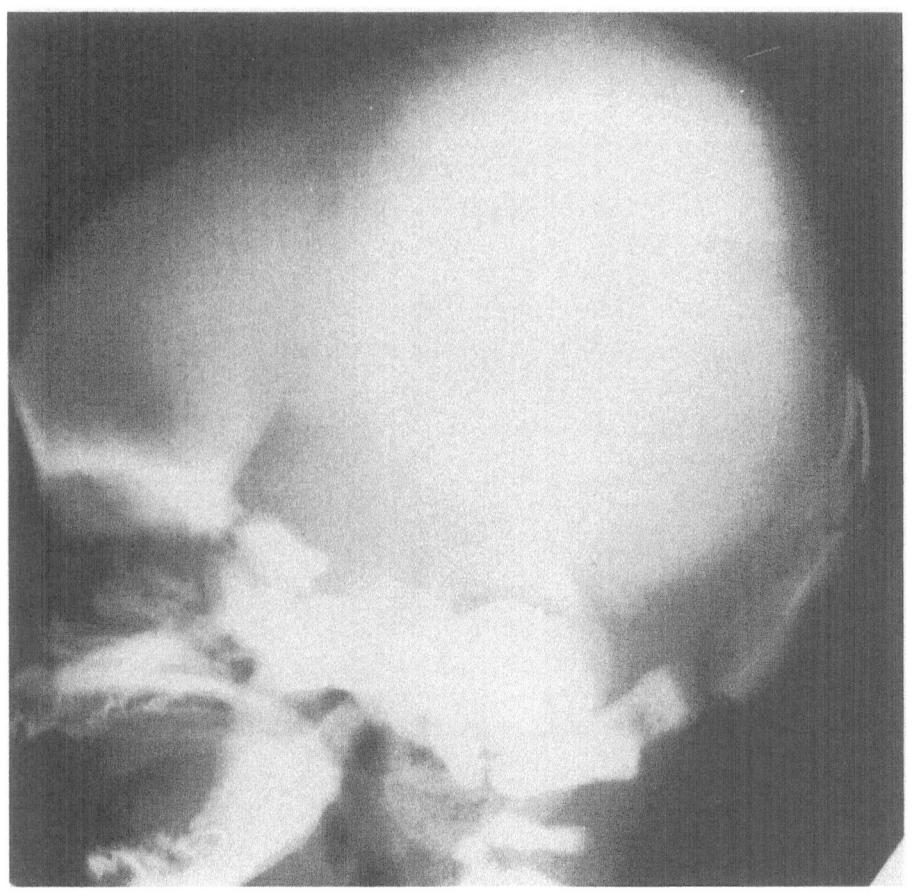

Fig. 8.3a

Cleidocranial Dysostosis

Case Report

Lateral Skull: There is defective ossification of the calvarium, with wide sutures and wormian bones. The bone density of the skull base is preserved and the occipital bone is slightly thickened.

Differential Diagnosis

1. Cleidocranial dysostosis is most likely in view of the preservation of skull base density, and a chest film should be obtained.
2. Osteogenesis imperfecta and hypophosphatasia cause more widespread loss of bone density.
3. Pyknodysostosis, hypothyroidism, and progeria should also be considered.

Supplementary Film

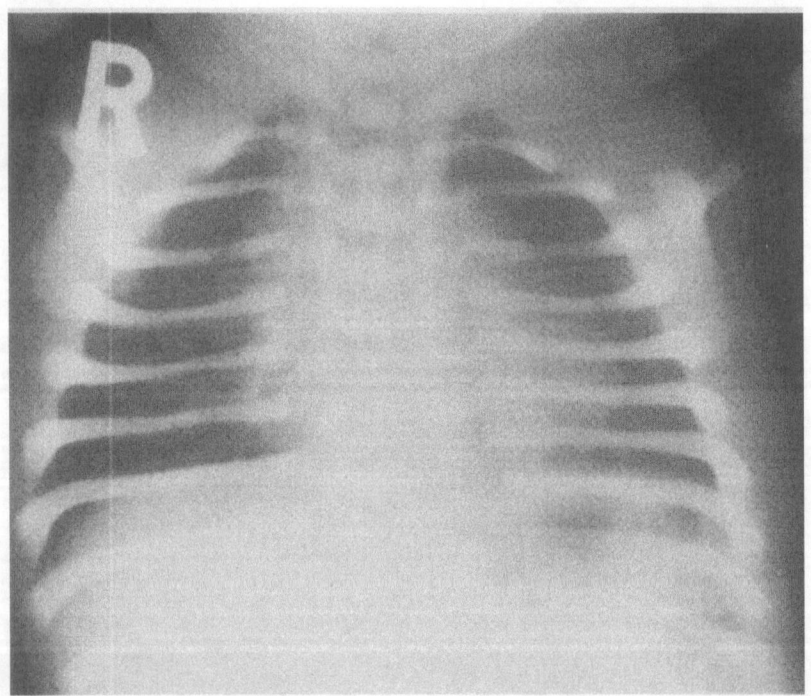

Fig. 8.3b

Chest Radiograph: *No clavicles can be seen, confirming cleidocranial dysostosis.*

134

Clinical Data

Female, 56 years old, with dysphagia and "PUO".

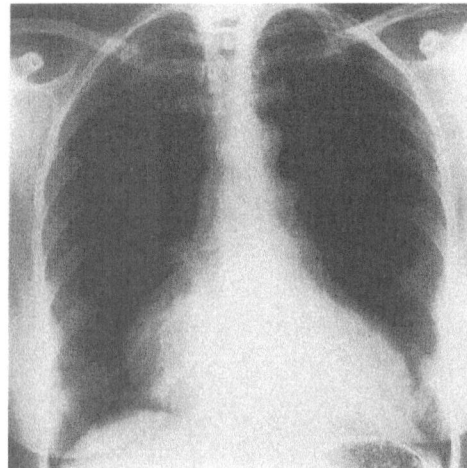

Fig. 8.4a

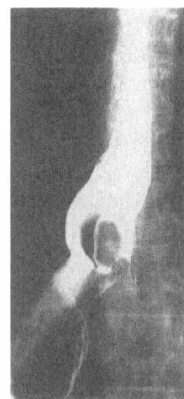

Fig. 8.4b

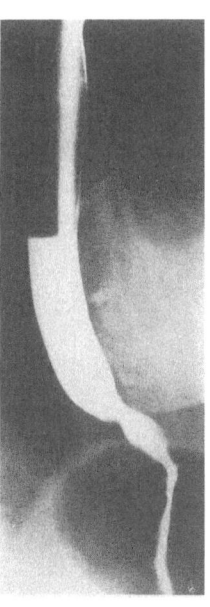

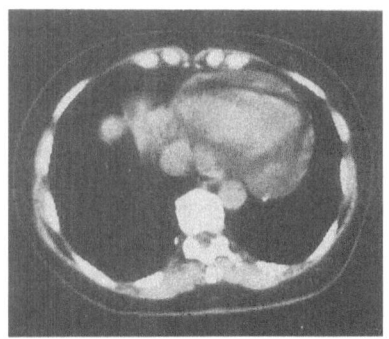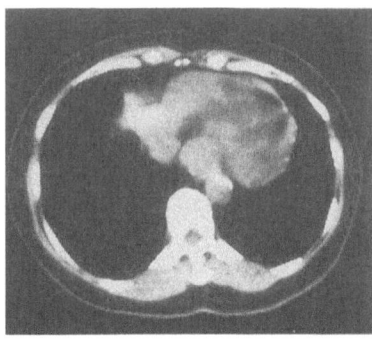

Fig. 8.4c

Lymphoma, With Pericardial Effusion

Case Report

PA Chest: The heart is enlarged and has a globular appearance without any obvious single chamber enlargement. The lungs are normal.

Barium Swallow: The prone film shows an apparent filling defect in the lower oesophagus which is smooth in outline. No filling defect is seen on the erect film, suggesting that the abnormality is due to an extrinsic mass compressing the oesophagus, rather than to an intrinsic lesion.

Chest CT (Enhanced): A pericardial effusion is present. There is a large, rounded, soft-tissue mass lying anterior to the oesophagus between the aorta and the IVC.

Differential Diagnosis

1. Lymphoma is the most likely cause for all the above appearances.
2. Other posterior mediastinal masses such as neurenteric cysts are possible, but do not cause a pericardial effusion.

Clinical Data

Male, 76 years old, with haematuria.

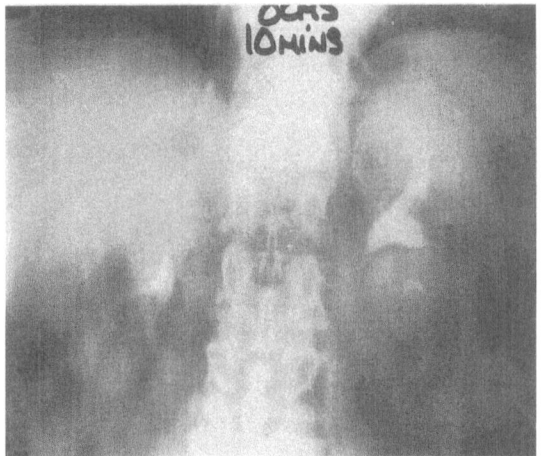

Fig. 8.5a

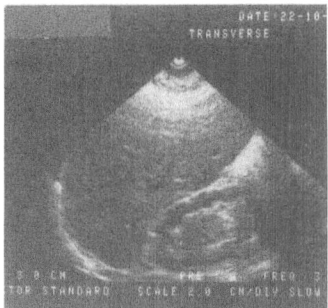

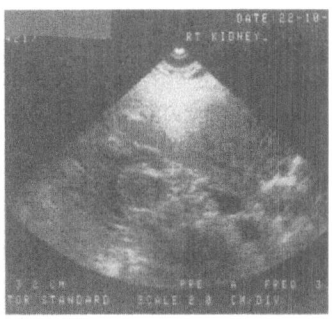

Fig. 8.5b

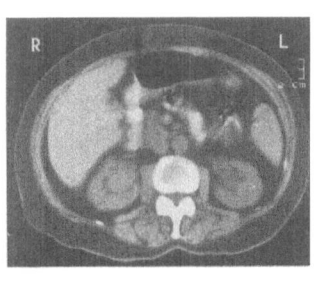
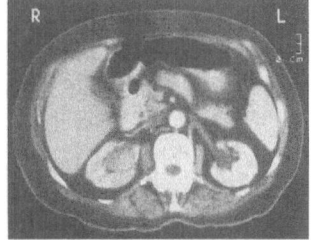

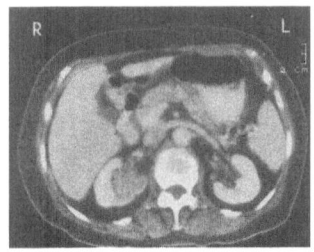
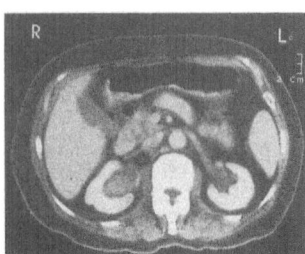

Fig. 8.5c

Renal Transitional-Cell Carcinoma

Case Report

IVU (Single Tomographic Film): The right upper and mid-pole calices are not visible and there is a curved indentation of the upper part of the right renal pelvis, suggesting that a mass is compressing this part of the right kidney. The right lower-pole calices, the left calices, and both renal outlines are normal.

Renal Ultrasound: A 3-cm long filling defect is seen within the right calices which does not cast an acoustic shadow and is therefore soft tissue. The mass extends into the renal pelvis.

Abdominal CT (Enhanced): There is a round soft-tissue filling defect lying in the right upper pelvicaliceal system.

Differential Diagnosis

1. Transitional-cell tumour.
2. Blood clot.

Both these possibilities are equally likely.

Clinical Data

Female, 64 years old, with breast carcinoma.

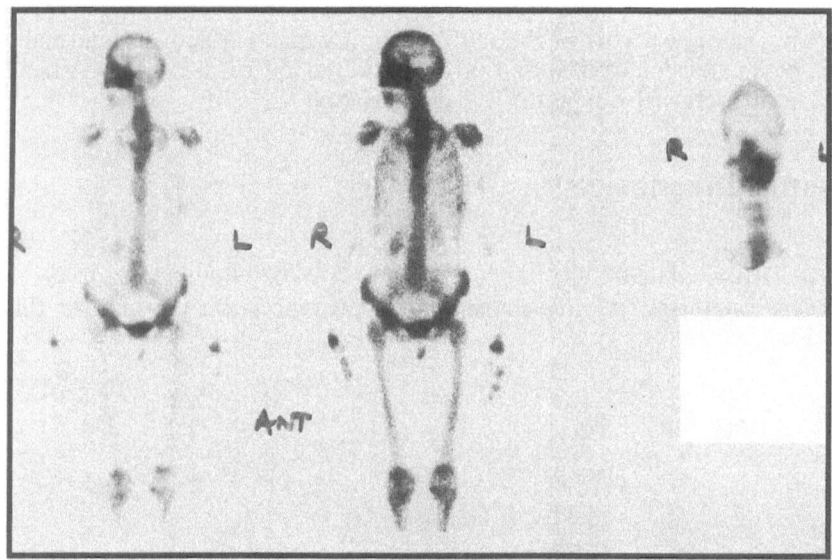

Fig. 8.6a

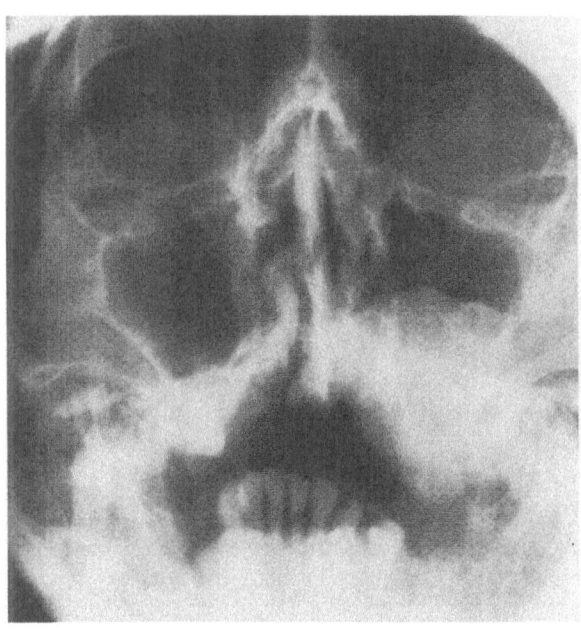

Fig. 8.6b

Fibrous Dysplasia of Left Maxillary Antrum

Case Report

Bone Scan: There is a localised area of increased uptake in the left mid-facial region at the level of the posterior orbital margin. No other abnormal area is seen.

Sinus View: The lower part of the left maxilla is expanded and is abnormally sclerotic, with extension of the sclerotic area into the left nasal cavities. A head CT scan would define the extent of the abnormality.

Differential Diagnosis

1. Fibrous dysplasia is the most likely cause for the above findings.
2. Possibly a single sclerotic metastasis, but appearances are unusual for this diagnosis.

Supplementary Film

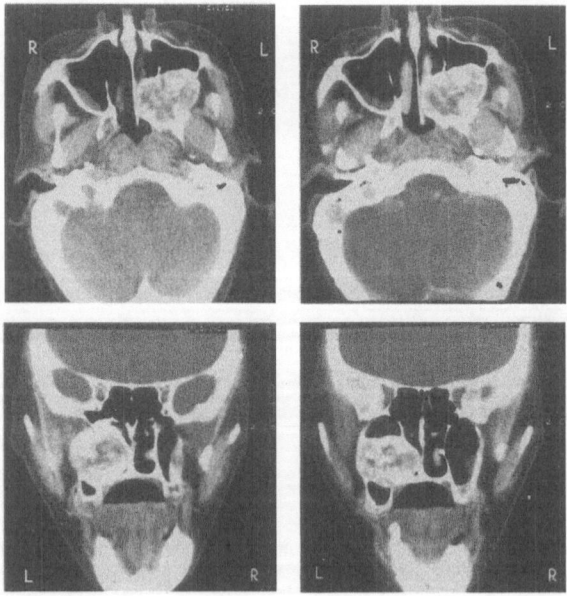

Fig. 8.6c

Axial and Coronal Sinus CT: *There is a large, well-defined mass in the left antrum, which is predominantly dense bone. The antral walls are thickened and sclerotic and the whole antrum is expanded. The appearances are those of fibrous dysplasia.*

Clinical Data

Male, 61 years old, with back pain.

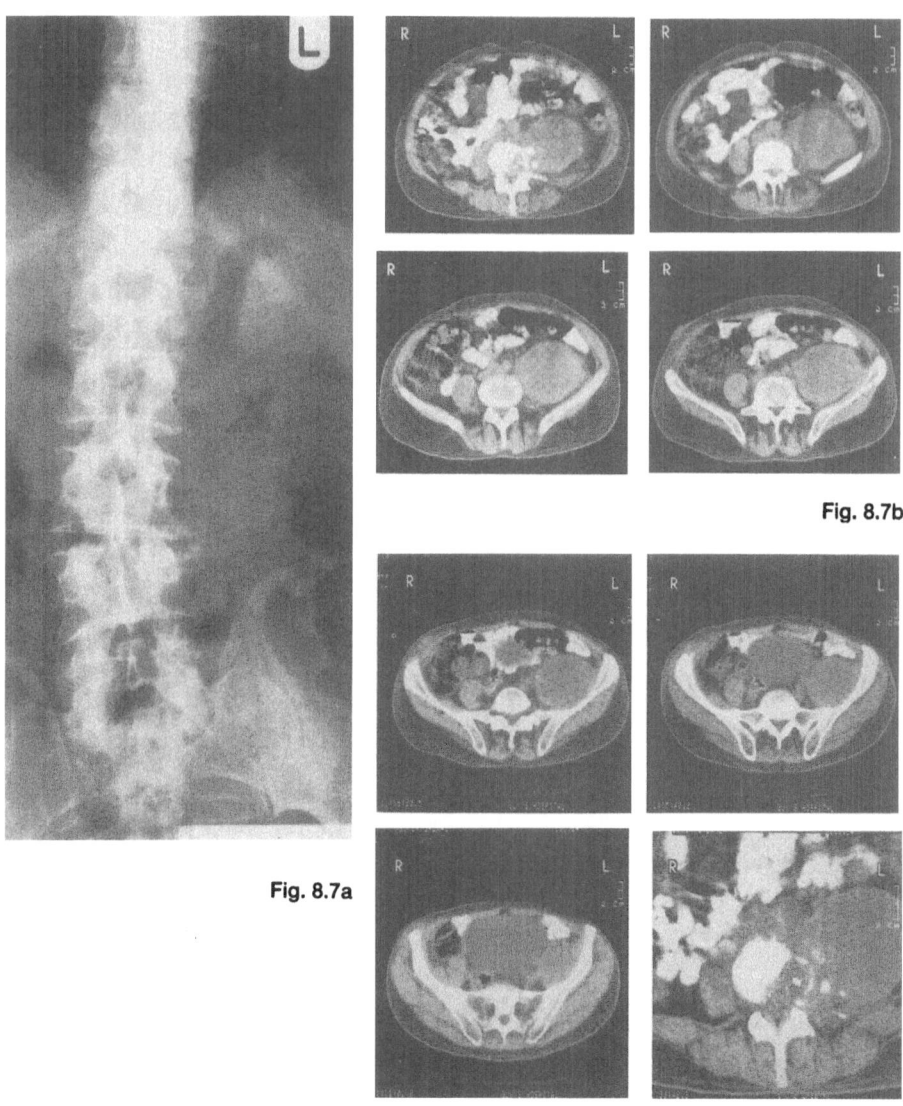

Fig. 8.7a

Fig. 8.7b

Fig. 8.7c

Tuberculous Psoas Abscess

Case Report

AP Lumbar Spine: There is a scoliosis concave to the left. The left psoas outline is displaced laterally in its lower part. The left L3–L4 disc space height is decreased with destruction of the left part of the L4 end-plate. The left L4 transverse process is ill defined and there is soft-tissue calcification lateral to the L3–L4 disc space. The left sacroiliac joint is normal.

CT Abdomen: The left side of the body of L4 is destroyed and there is a large associated soft-tissue mass on the left which has a low attenuation centre and a thick wall. This mass lies in the position of the left psoas muscle, and extends down within the iliopsoas in the pelvis. Poorly defined calcification is seen within this mass.

Differential Diagnosis

1. Tuberculous psoas abscess is the most likely diagnosis, particularly with the loss of disc space at L4.
2. A primary soft-tissue sarcoma involving the vertebrae is possible, but would not affect the vertebral disc.
3. Acute osteomyelitis could produce these appearances, but is less likely with the soft-tissue calcification.

Clinical Data

A 57-year-old male with chest pain.

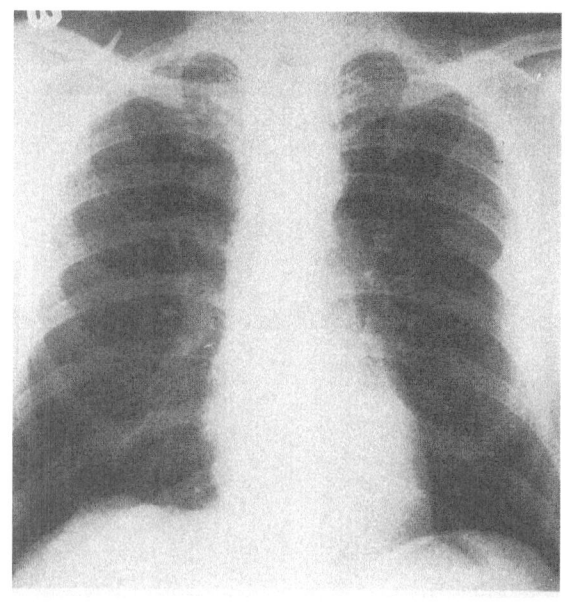

Fig. 8.8a

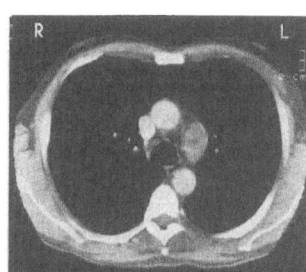
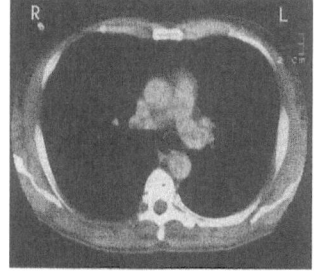
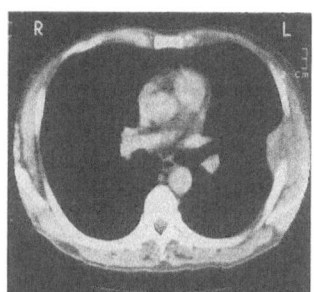
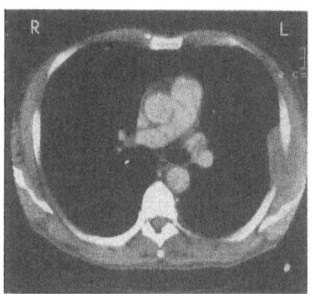

Fig. 8.8b

Bronchial Carcinoma

Case Report

PA Chest: There is a mass projected over the left hilar region which obscures the proximal parts of the hilar vessels. The descending aorta can be seen clearly, indicating that this mass probably lies in the middle mediastinum. A rounded, pleurally based mass is seen in the left mid-zone in the axillary line, with no associated rib lesion.

CT Chest: A large soft-tissue mass is present in the aortopulmonary window, which represents an enlarged lymph node. The pleural-based mass is well visualised and extends through the chest wall into the subscapular region.

Differential Diagnosis

1. Bronchial carcinoma with a pleural metastasis is the most likely cause for the above findings.
2. Metastasis from a distant primary is possible.
3. Lymphoma may also cause these appearances.

Examination 9

Clinical Data

A 45-year-old man whose wife noticed a lump on his head.

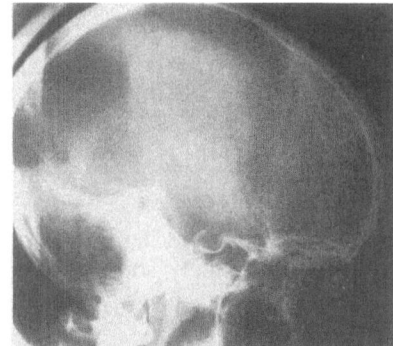

Fig. 9.1a

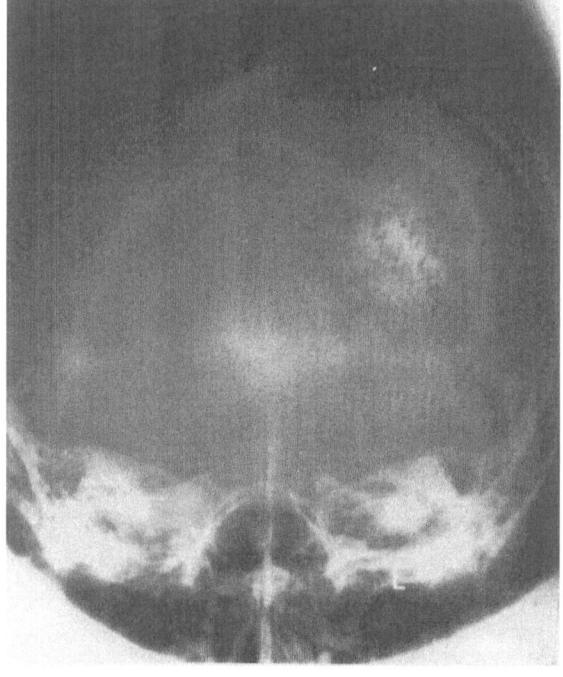

Fig. 9.1b

Skull Vault Haemangioma

Case Report

Lateral Skull: There is a semicircular area of patchy lucency in the anterior part of the frontal bone, which has a sclerotic rim and is associated with mild enlargement of the cranial vault at this site.

Townes View: The above abnormality lies on the left side and is roughly circular. The trabecular pattern is spoke-like in configuration.
   A tangential view would define the lesion more clearly.
   The appearances are of a slowly growing lesion.

Differential Diagnosis

1. Vault haemangioma is the most likely diagnosis and a head CT and external carotid angiogram would help confirm this.
2. A meningioma could possibly cause a similar appearance, but is usually more sclerotic.
3. An epidermoid cyst and chronic osteomyelitis cause lucent lesions with sclerotic margins in this age group, but they do not usually show such a variegated matrix.

Supplementary Film

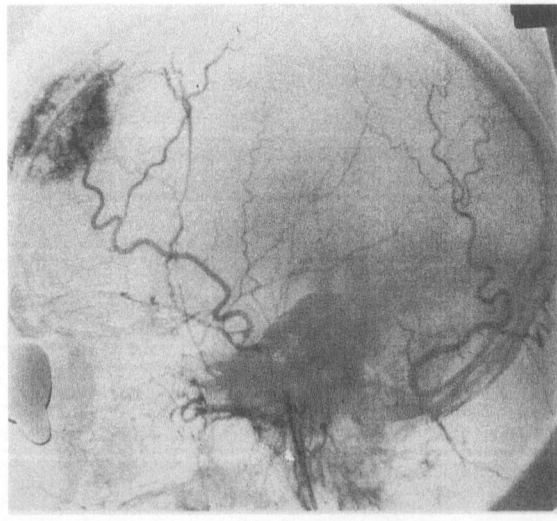

Fig. 9.1c

External Carotid Arteriogram: *This shows a prominent blush at the site of the described lesion, with a hypertrophied vascular supply.*

148

Clinical Data

A 53-year-old female with recurrent urinary-tract infections.

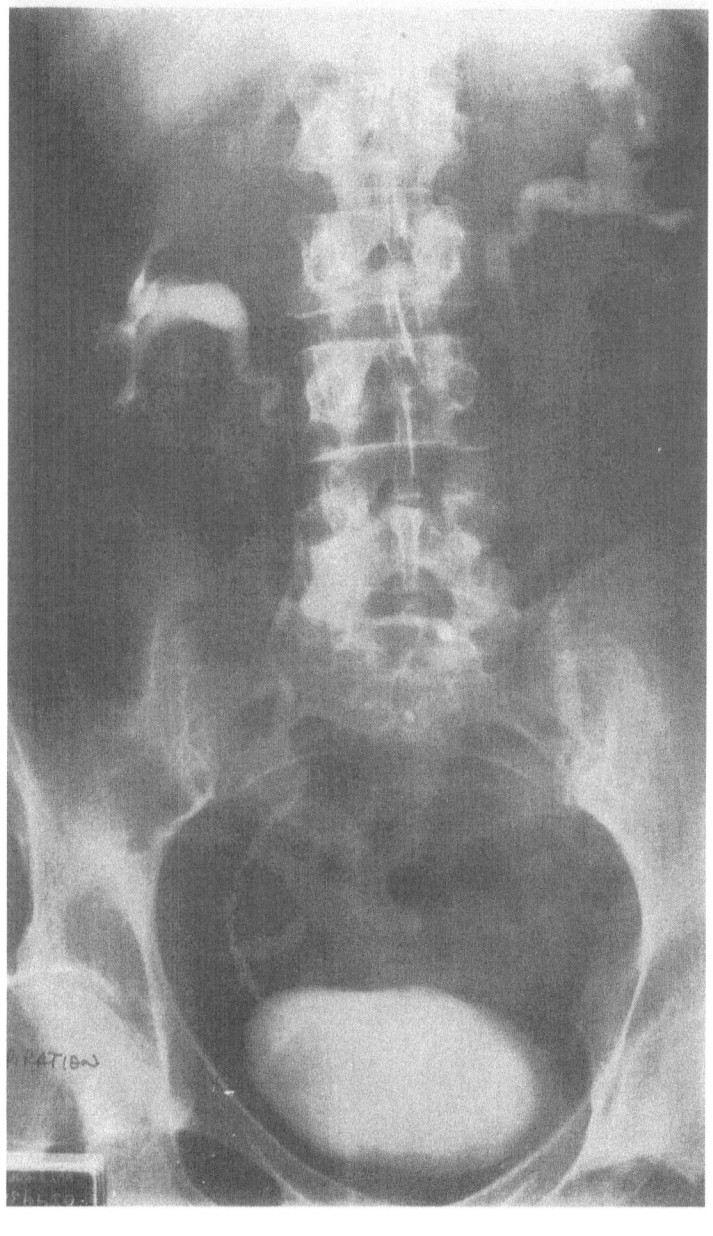

Fig. 9.2

Pyeloureteritis Cystica

Case Report

IVU (Single Full-Length Film): No control film is available. Myodil residue is seen over the sacral spine. The renal outlines are not clearly seen. The left pelvicaliceal system and ureter have multiple filling defects which are all less than 1 cm in diameter. The right system and the bladder appear normal.

Differential Diagnosis

1. With a background of chronic urinary-tract infections, pyeloureteritis cystica is most probable.
2. Multifocal transitional-cell carcinoma.
3. Blood clot; this is unlikely in the absence of haematuria.

Clinical Data

A 22-year-old female with dyspnoea.

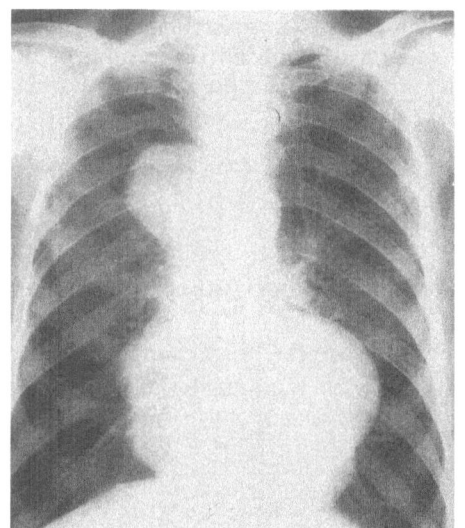

Fig. 9.3a

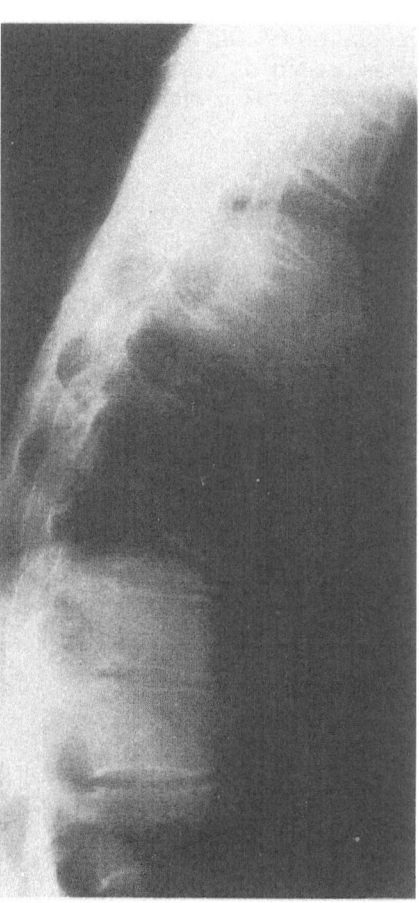

Fig. 9.3c

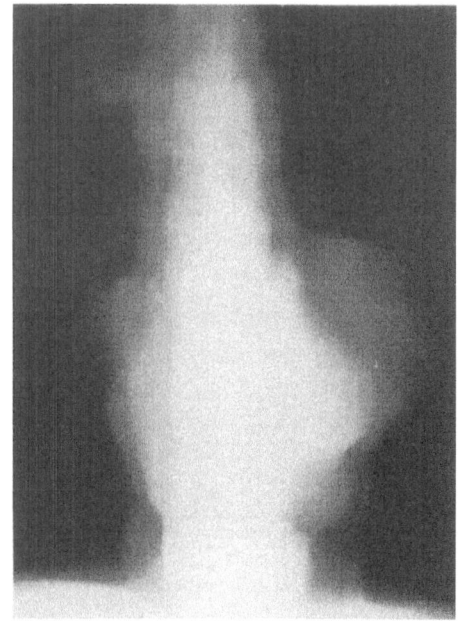

Fig. 9.3b

Thalassaemia Major

Case Report

PA Chest: There are several large, well-defined, rounded masses projected on either side of the spine. They are probably posterior in position, since the heart borders are well seen.

Penetrated PA and Lateral: These confirm the position of the masses. There is no associated rib erosion, but the ribs show a generalised coarsening of the trabecular pattern with widening of the medullary component and cortical thinning.

Differential Diagnosis

1. The appearances are typical of thalassaemia major with extramedullary haematopoiesis.
2. Neurofibromata are possible but would not cause the bony changes.
3. Multiple lymph-node enlargement is also possible but again does not explain the bone changes.

Clinical Data

A 72-year-old male with diplopia.

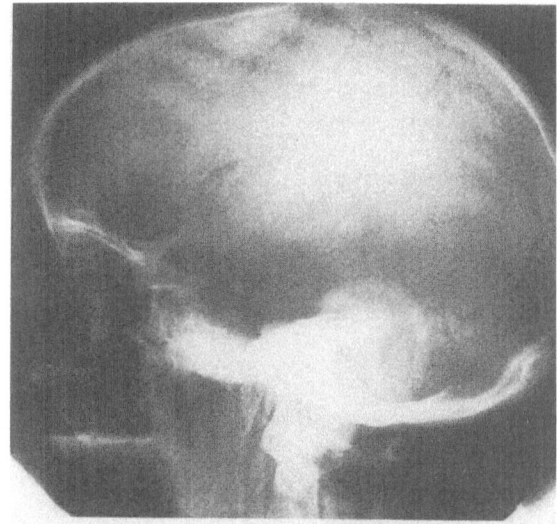

Fig. 9.4a

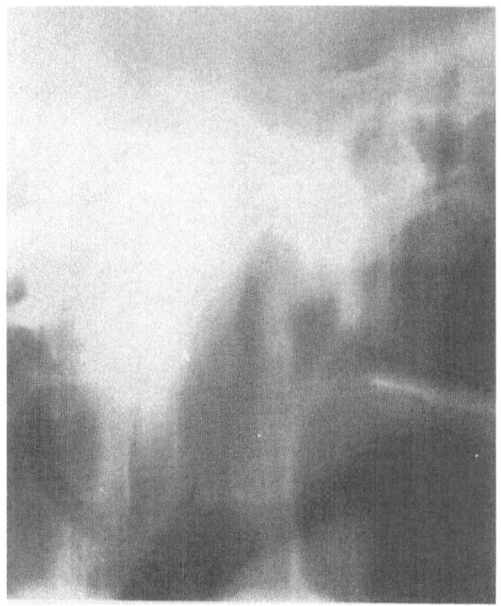

Fig. 9.4b

Nasopharyngeal Carcinoma

Case Report

Lateral Skull and Plain Tomogram of Pharynx: There is destruction of the clivus and the dorsum sellae, with poor definition of the floor of the sella, suggesting that this area is also destroyed.

A huge soft-tissue mass is visible in the nasopharyngeal region, displacing the airway anteriorly. The cervical vertebrae are not involved.

The rest of the skull is normal.

Differential Diagnosis

1. Nasopharyngeal carcinoma.
2. Chordoma.

Both these conditions could cause the above findings, but chordoma is less likely to occur in this age group.

A CT scan would be useful to define the extent of the abnormality more clearly.

Supplementary Film

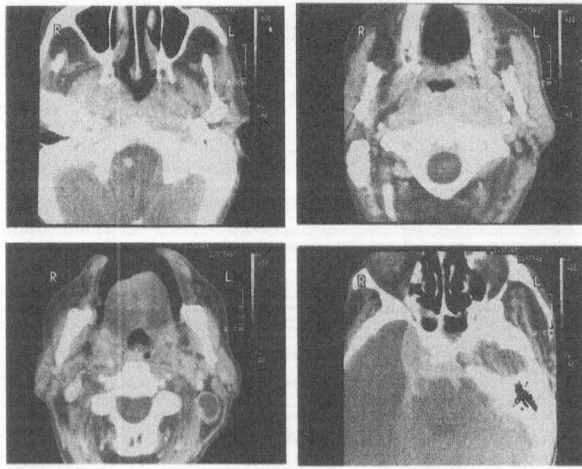

Fig. 9.4c

CT of Upper Neck: *There is a large nasopharyngeal mass displacing the airway and extending intracranially to involve the right cavernous sinus. There is a 1.5-cm diameter lymph node posterior to the left parotid which has an enhancing rim and a low-attenuation centre. The appearance of the lymph node is highly suggestive of a squamous-cell metastasis, confirming the diagnosis of nasopharyngeal carcinoma.*

Clinical Data

A 60-year-old female with back pain.

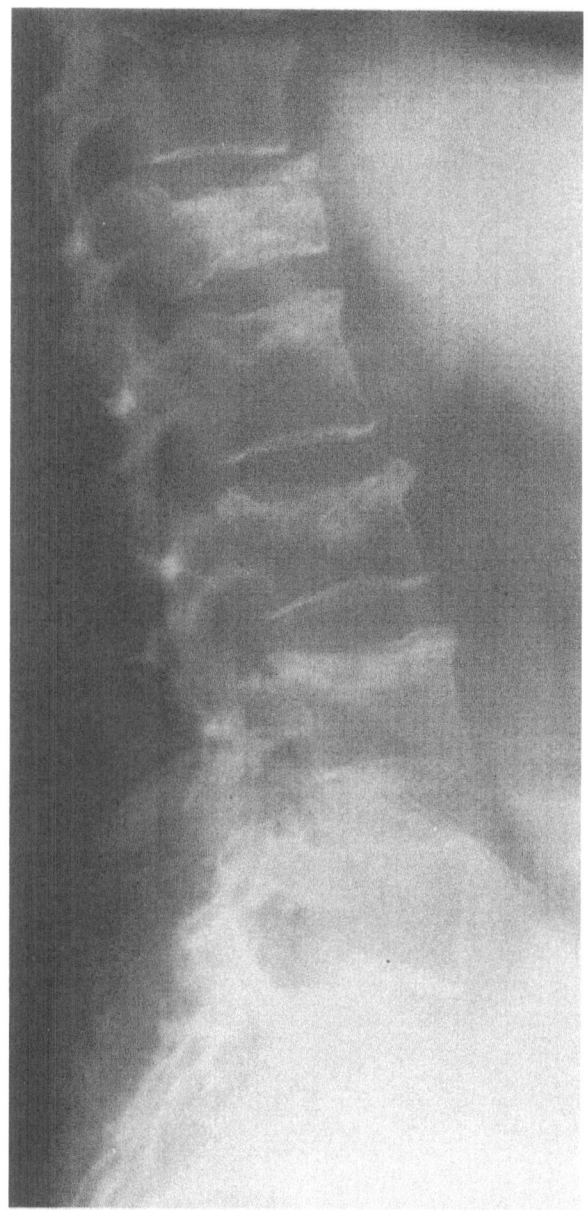

Fig. 9.5

Cushing's Syndrome

Case Report

Lateral Lumbar Spine: The bones are osteopenic and the vertebral bodies are biconcave with marked collapse of the body of L1. There are horizontal sclerotic bands adjacent to the vertebral end-plates, which represent excessive callus formation.

Diagnosis

The appearances are typical of steroid-induced osteoporosis, either Cushing's disease or Cushing's syndrome.

Clinical Data

A 42-year-old female with facial lumps.

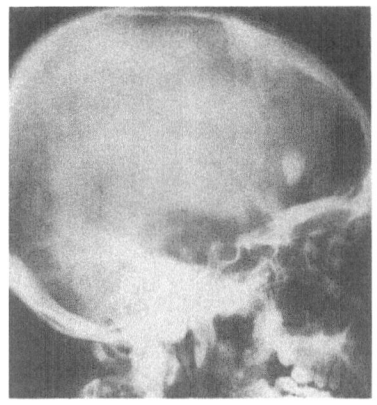

Fig. 9.6a

Fig. 9.6b

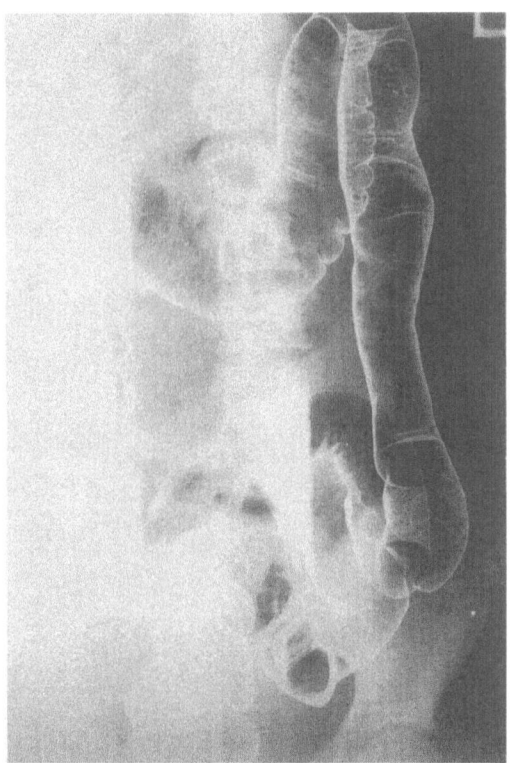

Fig. 9.6c

Gardner's Syndrome

Case Report

Skull and Mandible: A dense, sclerotic, well-defined lesion is seen in the frontal region, with a further lobulated sclerotic lesion protruding from the right mandibular angle. There is at least one further lesion in the maxilla. The appearances are those of multiple osteomata.

Barium Enema: There are numerous variable-sized rounded lesions throughout the large bowel showing the features of multiple polyps. The largest polyp is 5 mm in diameter, and no features of colonic malignancy are seen.

Diagnosis

The combination of multiple large-bowel polyps and osteomata is diagnostic of Gardner's syndrome.

Clinical Data

Male, 65 years old, with chronic dyspnoea.

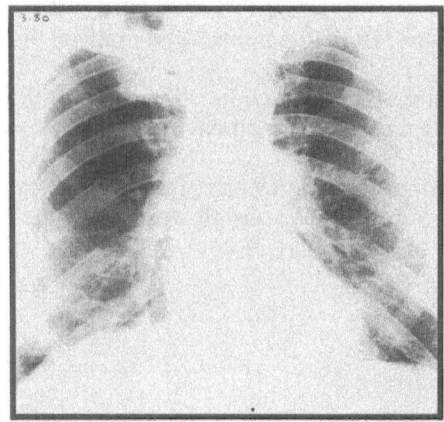

Fig. 9.7a

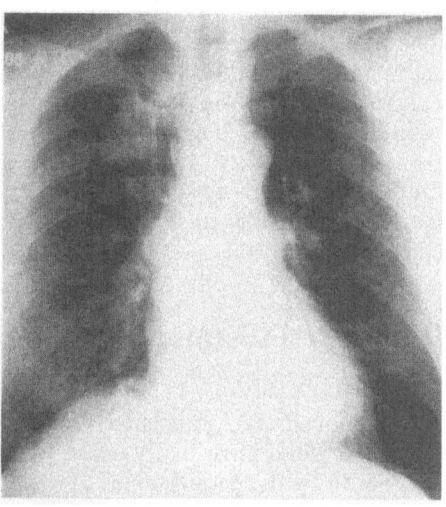

Fig. 9.7b

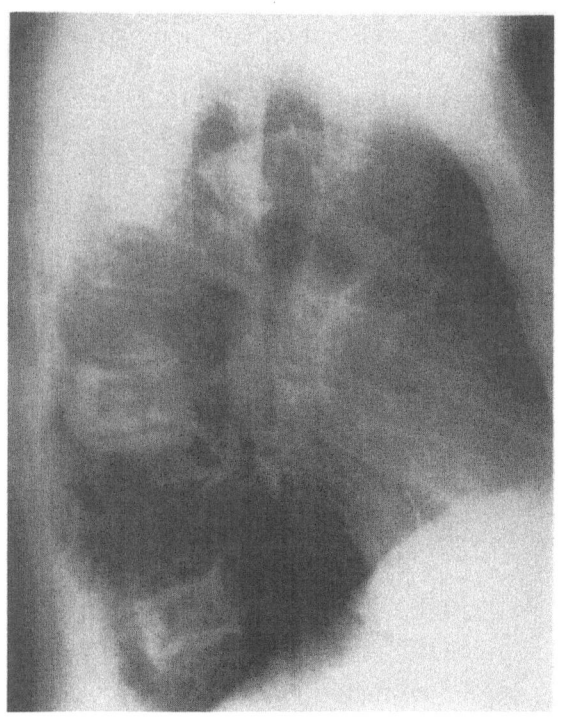

Fig. 9.7c

Pneumoconiosis with Progressive Massive Fibrosis and Paget's Disease of the Thoracic Spine

Case Report

Chest PA 1980–1985: The 1980 radiograph shows multiple, widespread, small nodules of uniform density, more obvious at the bases. There is also a 3-cm diameter circular mass in the right upper zone. Both hila appear quite dense, but show no calcification.

Five years later the nodular background pattern is much less obvious and the right apical lesion is unchanged in size but now has a small cavity within it.

Lateral Film: This does not show the upper zone mass clearly, but there are at least three expanded, dense, sclerotic vertebral bodies visible with coarsened trabecular patterns.

Differential Diagnosis

1. Pneumoconiosis, particularly silicosis, with associated progressive massive fibrosis is the most likely cause for the above changes over the 5-year period covered by the radiographs.
2. Sarcoidosis could produce all these changes.
3. Metastases and lymphoma are less likely in view of the lack of progression of the upper lobe mass over 5 years.
4. The appearances of the thoracic vertebrae are those of Paget's disease.

Clinical Data

A 5-year-old girl with a painful, swollen finger.

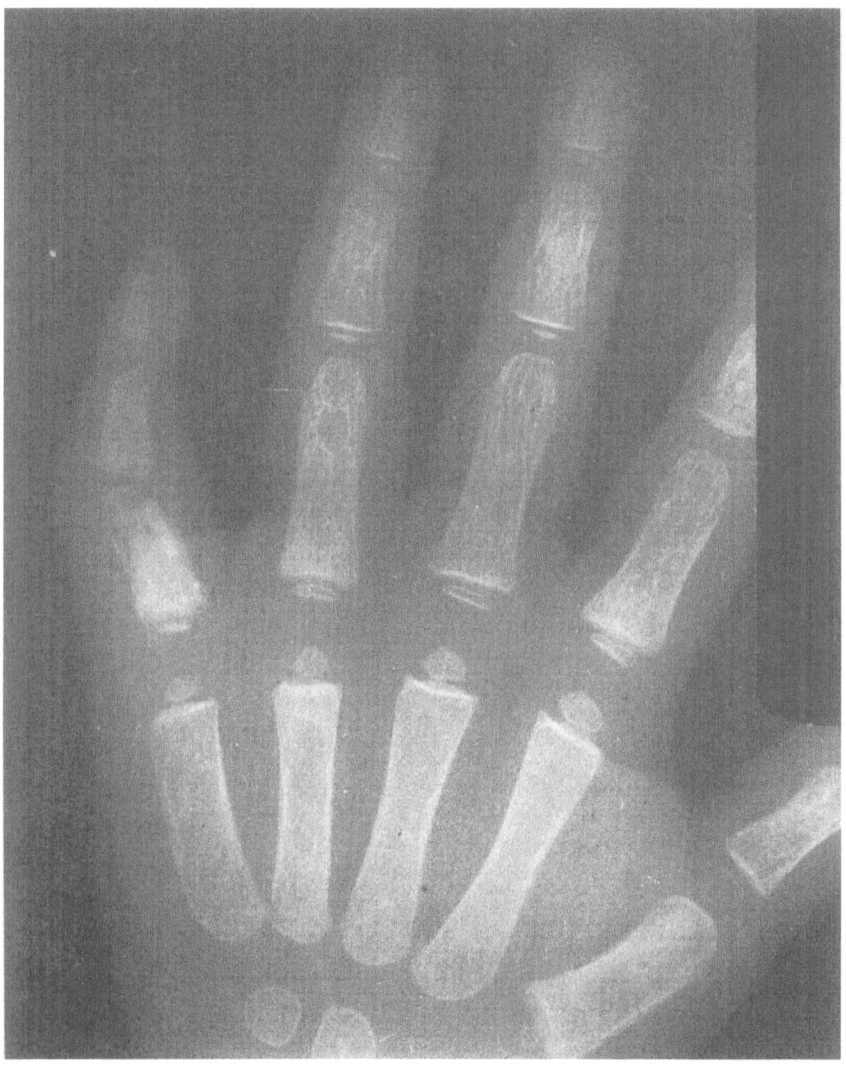

Fig. 9.8

Sickle-Cell Dactylitis

Case Report

Hand: The proximal phalanx of the little finger shows a non-homogeneous increase in density, and the outline is irregular in comparison with the other phalanges. There is bone destruction and a periosteal reaction in the distal part of this phalanx, with associated soft-tissue swelling. The other digits show generalised osteopenia.

Differential Diagnosis

1. Sickle-cell dactylitis, possibly complicated by osteomyelitis. What is the patient's racial origin?
2. Primary osteomyelitis, in particular tuberculosis, should be considered.

Examination 10

Examination 10

Clinical Data

Male, 75 years old, with severe abdominal pain.

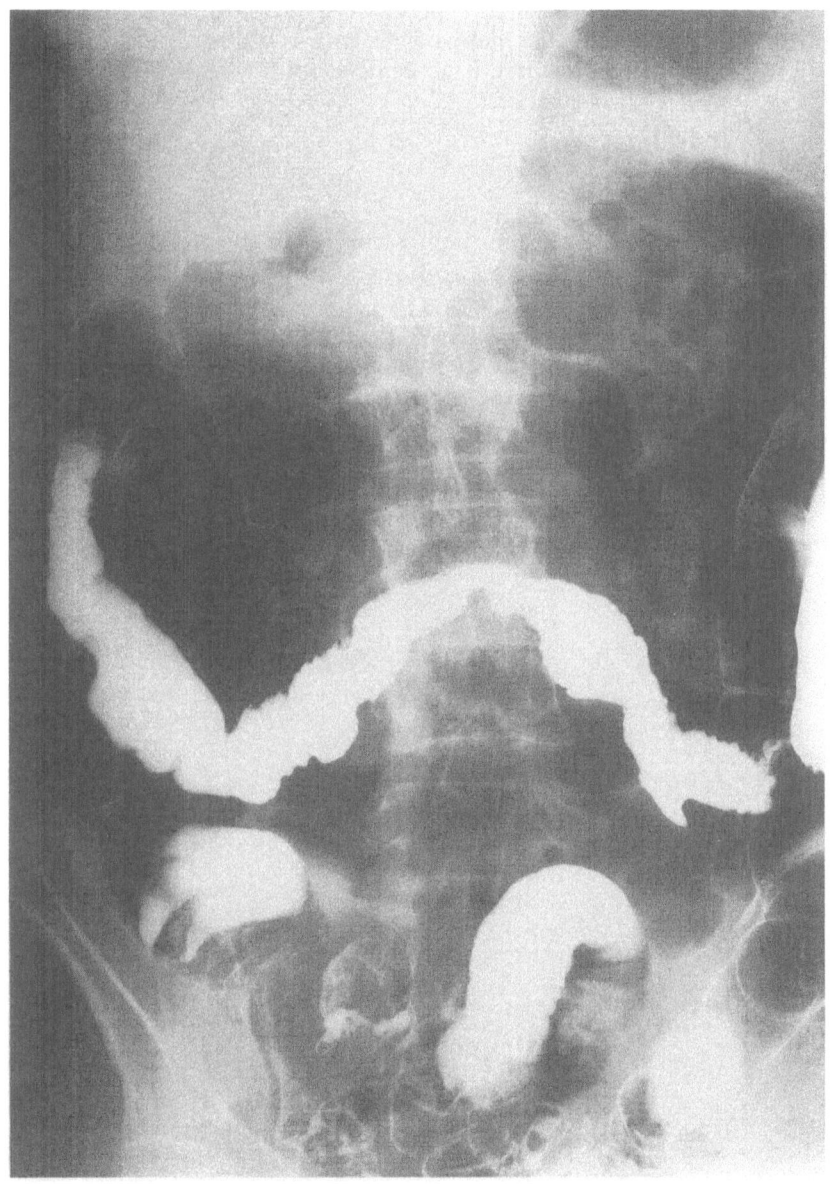

Fig. 10.1

Ischaemic Colitis

Case Report

Barium Enema: The transverse colon shows "thumbprinting" pattern of the mucosa with pseudosacculation. There is in addition apparent narrowing of the distal sigmoid colon with probable stricture formation. Further views are indicated to confirm these findings. There are also linear lucencies in the periphery of the liver, and further views of this area would confirm this finding. The small bowel is dilated but there is no evidence of bowel perforation.

Differential Diagnosis

1. The appearances suggest ischaemic colitis with gas in the portal vein.
2. Severe inflammatory bowel disease is also possible but is unlikely to cause portal-vein gas.

The presence of portal-vein gas in an adult has a very poor prognosis and urgent treatment is indicated.

Clinical Data

A 75-year-old male with raised alkaline phosphatase.

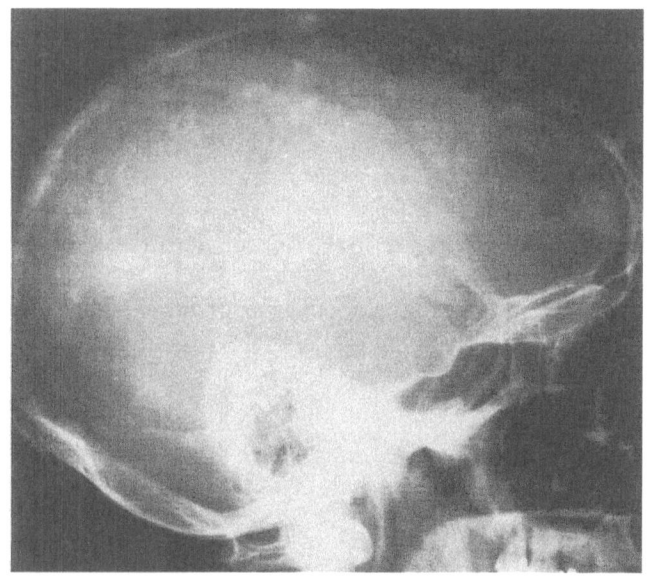

Fig. 10.2a

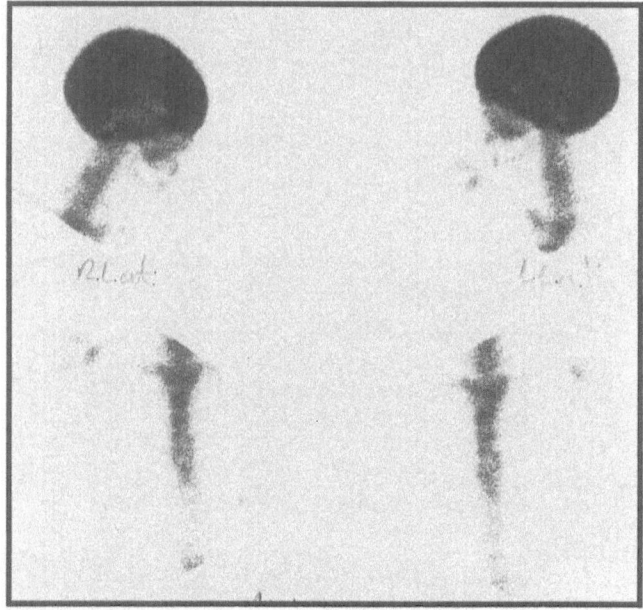

Fig. 10.2b

Paget's Disease of the Skull

Case Report

Lateral Skull: There is a diffuse abnormality of the skull vault. The table is thickened and there are multiple poorly defined sclerotic areas 1–2 cm in diameter. Several relatively well-defined lucent areas are also visible in the skull vault. The skull base is normal and the inner table is preserved.

Bone Scan: Massive increased uptake is present throughout the calvarium, with normal uptake in the rest of the visible skeleton.

Differential Diagnosis

1. The appearances are those of Paget's disease in the mixed sclerotic and lucent phase.
2. Metastases do not normally produce this combination of findings and are therefore very unlikely.

Clinical Data

A 19-year-old male with pain in the right leg.

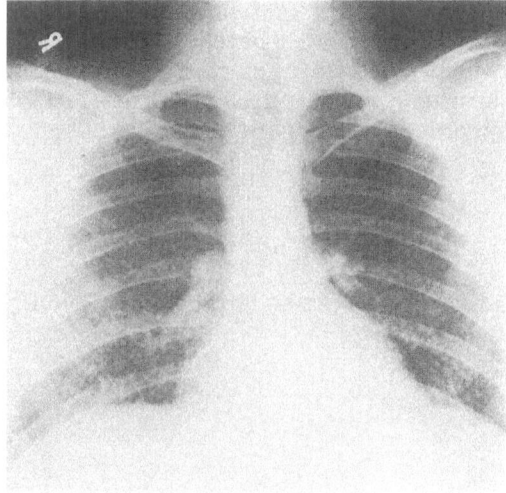

Fig. 10.3a

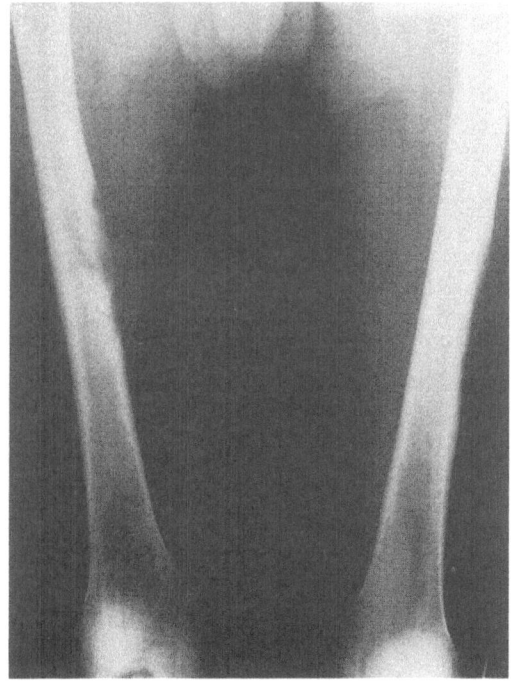

Fig. 10.3b

Histiocytosis X

Case Report

Chest Radiograph: There is widespread reticular density throughout both lungs, cystic in places and forming a "honeycomb" pattern. No hilar enlargement, volume loss, or pleural abnormality is seen.

Both Femora: A large, well-defined lytic lesion is present in the right mid-femoral shaft. This is expanding the cortex and shows a bevelled "hole within a hole" appearance. A smaller, well-defined scalloped lesion is seen in the medial cortex distally on the right, with distal osteopenia. A similar lesion is present in the mid-shaft of the left femur, with cortical expansion proximally.

Diagnosis

The combination of characteristic chest abnormalities and bone lesions is typical of histiocytosis X.

Clinical Data

A 12-year-old girl with headaches.

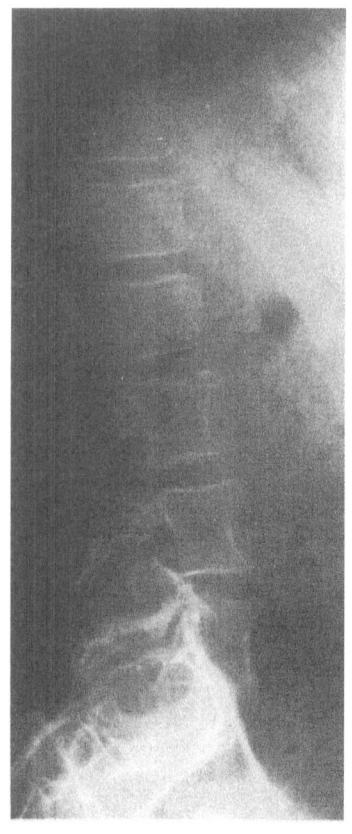

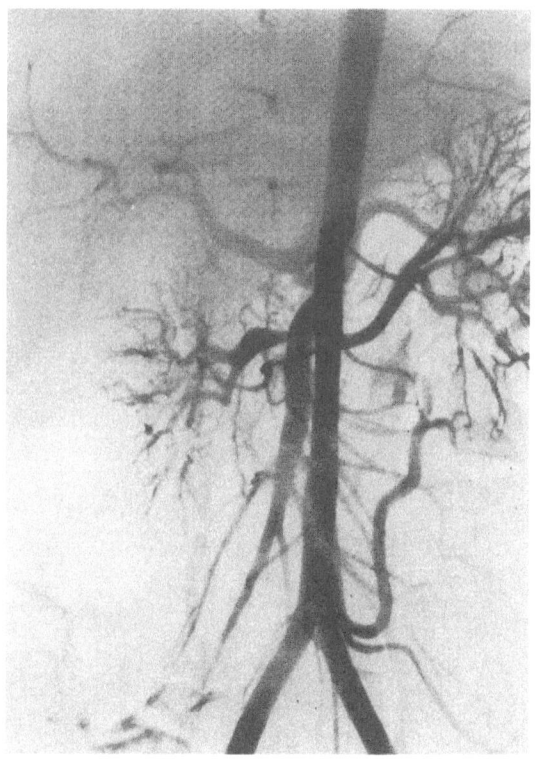

Fig. 10.4a

Fig. 10.4b

Neurofibromatosis

Case Report

Lateral Lumbar Spine: There is prominent anterior and posterior scalloping of the bodies of the lower lumbar vertebrae, and the lower lumbar and upper sacral intervertebral foraminae are markedly enlarged. Prominent angulation is present at the L5–S1 junction.

Abdominal Aortogram: There is a marked concentric stenosis of the right main renal artery, the left renal artery is normal. The proximal abdominal aorta is also narrowed, suggesting an abdominal coarctation.

Differential Diagnosis

1. Neurofibromatosis causes vertebral scalloping and is a well-described cause of renal-artery stenosis in childhood.
2. Other causes of posterior vertebral scalloping such as Marfan's syndrome, Morquio's disease, achondroplasia etc., are not associated with renal-artery stenosis.

Clinical Data

Female, 62 years old, with chronic shortness of breath.

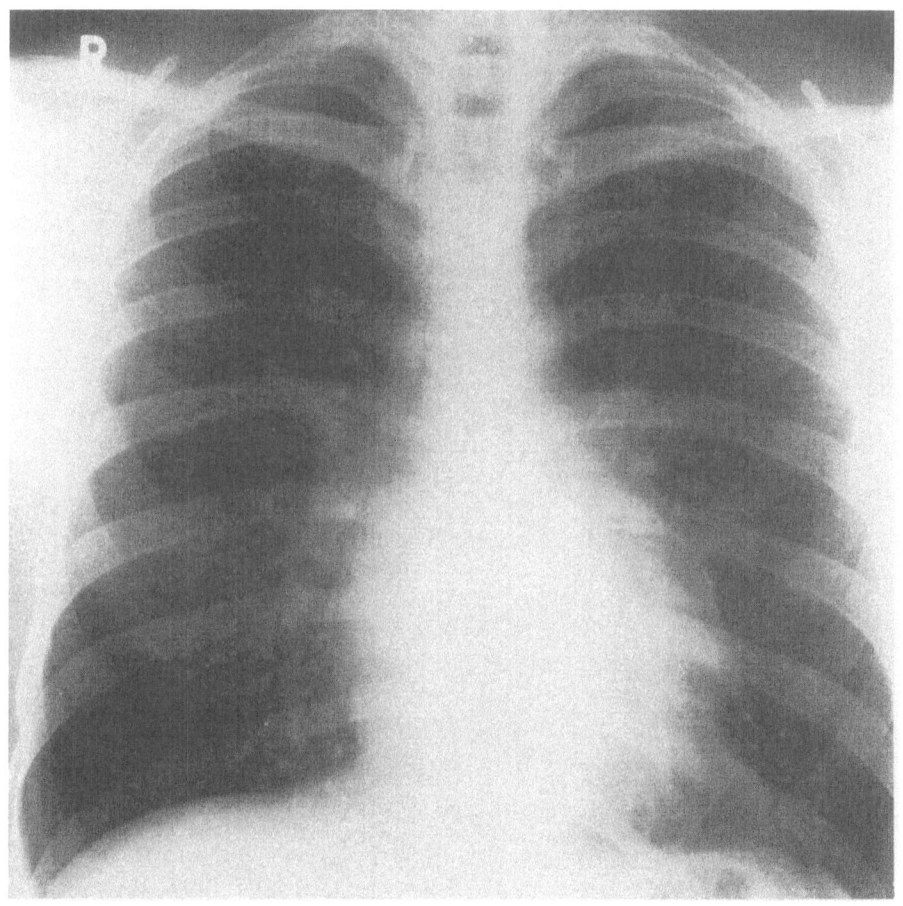

Fig. 10.5

173

Chronic Mitral Stenosis

Case Report

PA Chest: The heart is not enlarged, but there is a prominent bulge of the left heart border in the region of the left atrial appendage. The right heart border shows a double outline, but the carina is not adequately visualised. These findings suggest left atrial enlargement. There are Kerley B lines at the bases, together with upper-lobe blood diversion. Numerous small, circular calcific nodules (less than 0.5 cm in diameter) are seen in all areas.

Diagnosis

The appearances are of chronic mitral-valve stenosis with raised pulmonary venous pressure and ossific pulmonary nodules.

Clinical Data

A 24-year-old female with dysuria.

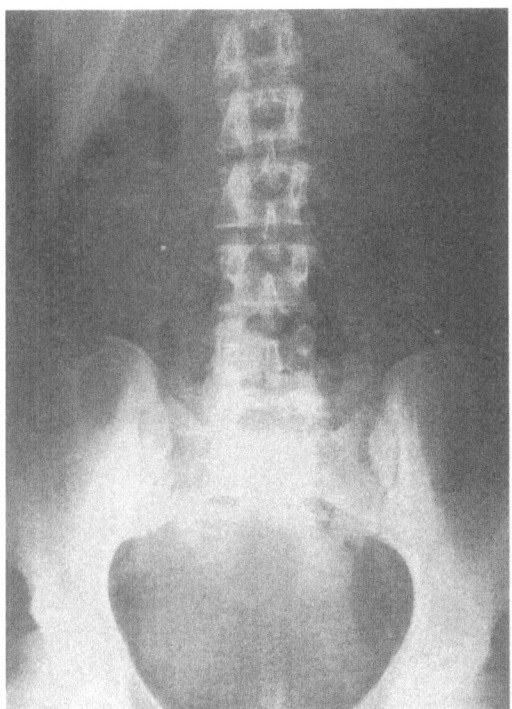

Fig. 10.6a

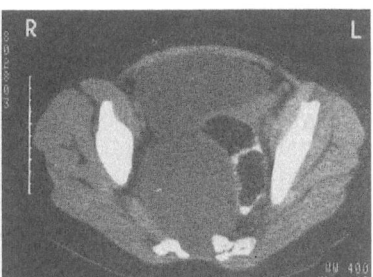

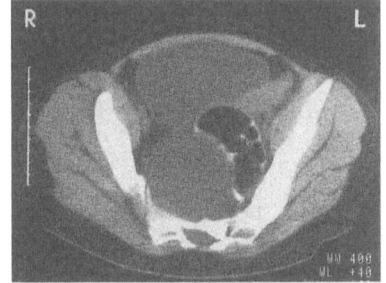

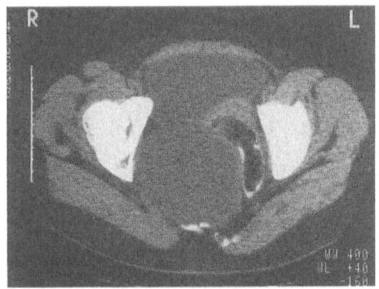

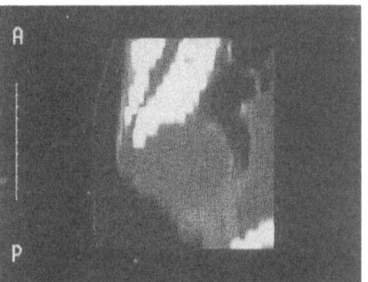

Fig. 10.6b

Anterior Sacral Meningocele

Case Report

AP Lumbar Spine: A well-defined circular defect which has a sclerotic margin is seen in the midline of the sacrum. The fat planes in the pelvis are displaced laterally by a midline soft-tissue mass, and no gas is visible in the pelvis. A slight scoliosis, concave to the right, is present, and the lamina of S1 is incompletely fused.

CT Pelvis: There is a midline anterior defect in the sacral spine, associated with a large well-defined soft-tissue mass which is close to water density. The rectum is displaced laterally and the mass extends into the spinal canal. The sagittal reformat confirms that the mass is separate from the bladder.

Differential Diagnosis

1. Anterior sacral meningocele is the most likely cause for these appearances.
2. Chordoma or metastases are possible, but are usually less well defined and would not be of water density.
3. A primary bone lesion such as a giant-cell tumour is less likely with such a large soft-tissue mass of relatively low attenuation.

Clinical Data

A 70-year-old male with acute renal failure.

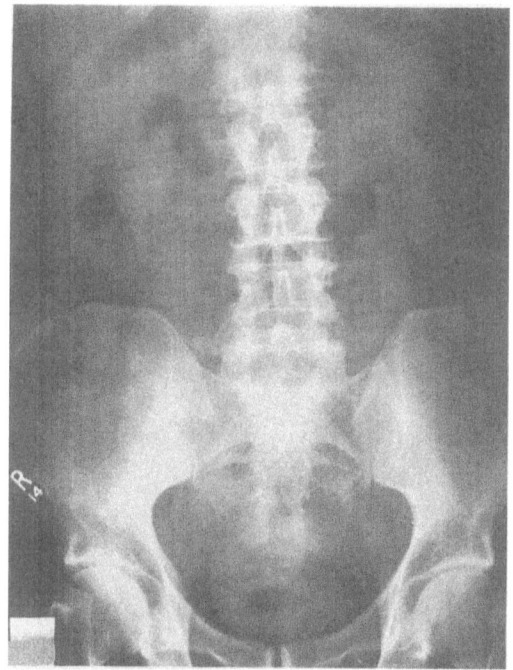

Fig. 10.7a

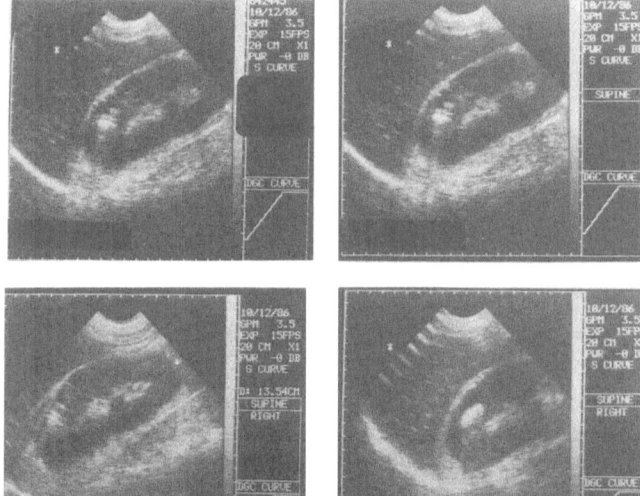

Fig. 10.7b

Renal Abscess Containing Gas

Case Report

Plain Abdomen: There is a crescentic gas shadow in the right upper quadrant projected over the right renal area. This gas shadow appears to follow the contour of the right upper renal margin and may represent gas within the renal parenchyma. No gas is seen in either the bladder or the ureters.

Renal Ultrasound: An echogenic mass is present in the upper pole of the right kidney, with a surrounding area of sonolucency replacing the normal cortical outlines. There is some poorly defined variable acoustic shadowing associated with this lesion. The pelvicaliceal systems are normal.

The appearances are those of gas in the upper pole of the right kidney within a mass lesion. A CT scan would be useful to define the abnormality more clearly.

Differential Diagnosis

1. Infection with gas-forming organisms is the most likely cause, probably associated with a renal abscess.
2. A fistula between the gut and urinary tract is also possible, but the distribution of the gas is unusual for this.

Supplementary Film

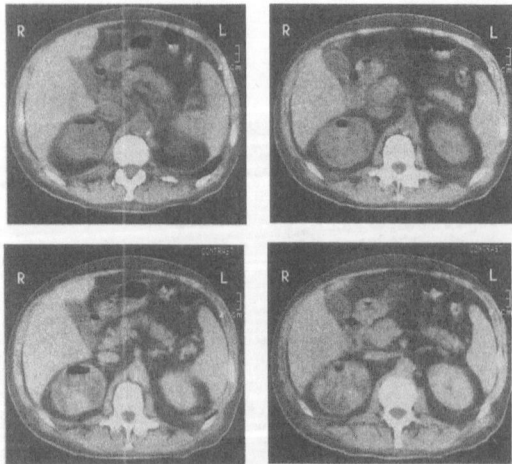

Fig. 10.7c

Abdominal CT (Pre- and Post-Contrast): *There is gas in the right kidney which is enlarged and contains areas of poorly defined low attenuation. The right kidney excretes no contrast but there is some parenchymal enhancement around the gas. The appearances are in keeping with a gas-containing renal abscess.*

Clinical Data

A 2-year-old girl.

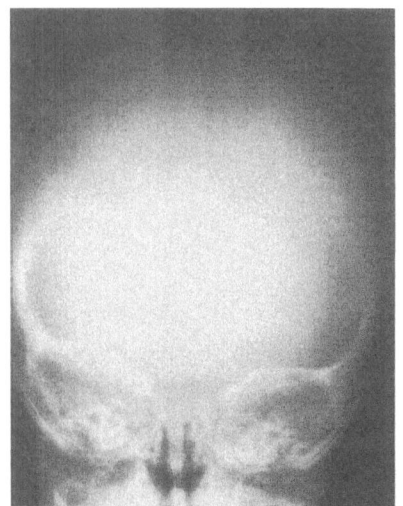

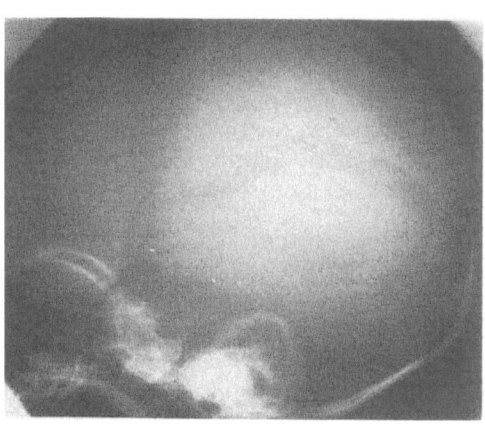

Fig. 10.8b

Fig. 10.8a

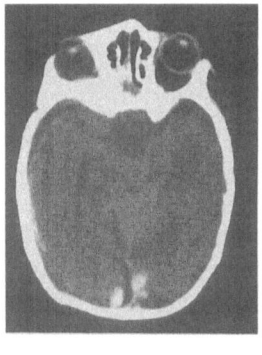

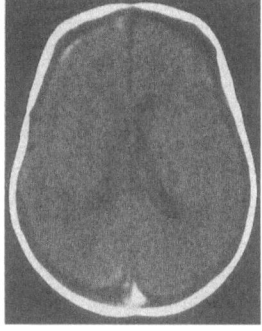

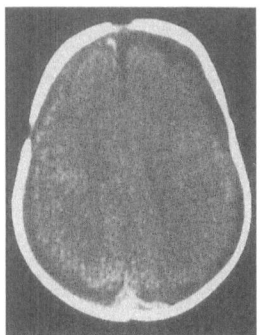

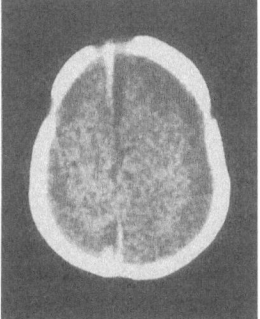

Fig. 10.8c

Non-Accidental Injury

Case Report

AP and Lateral Skull: There is a long, wide, horizontal lucency extending from the coronal suture anteriorly to the torcula posteriorly, which represents separated fracture segments. The sutures are of normal width and no wormian bones are seen.

CT Head: There are areas of increased attenuation adjacent to the torcula, which probably represent fresh blood. Bilateral elliptical areas of low attenuation are present, surrounding both cerebral hemispheres and extending into the interhemispheric fissure. The appearances are of bilateral, extensive, chronic subdural haematomas with fresh haemorrhage posteriorly.

Diagnosis

The above findings suggest non-accidental injury and a skeletal survey would be useful to detect the presence of other fractures.